D1127578

SQUIRRELS OF NORTH AMERICA

BY DORCAS MACCLINTOCK

ILLUSTRATIONS BY WALTER FERGUSON

A California Academy of Sciences Book

VNR VAN NOSTRAND REINHOLD COMPANY
NEW YORK CINCINNATI TORONTO LONDON MELBOURNE

Van Nostrand Reinhold Company Regional Offices:
New York • Cincinnati • Chicago • Millbrae • Dallas

Van Nostrand Reinhold Company Foreign Offices:
London • Toronto • Melbourne

Library of Congress Catalog Card Number 74-110059

 Manufactured in the United States of America.

Designed by Howard Burns
Printed and bound by Mahoney and Roese, Inc.

Published by Van Nostrand Reinhold Company,
450 West 33rd Street, New York, N. Y. 10001

Published simultaneously in Canada by
Van Nostrand Reinhold Ltd.

16 15 14 13 12 11 10 9 8 7 6 5 4 3 2 1

CONTENTS

II THE TREE SQUIRRELS

III THE GLIDERS

LIST OF ILLUSTRATIONS

ACKNOWLEDGMENTS

In a book such as this the list of references should come under Acknowledgments, for I have drawn heavily upon the observations of many scientists and naturalists. Walter Ferguson is to be thanked for his illustrations that capture so well the essence of being a squirrel. I am grateful to Dr. Robert T. Orr, of the California Academy of Sciences, and to Dr. George E. Lindsay, its director, for their suggestion that I do a book on the squirrels of North America. Thanks are due my parents, Mr. and Mrs. James T. Eason, who often have supported, always tolerated, my interest in the natural world and, especially, its mammals. My husband, Copeland MacClintock, offered much encouragement and toted many *volumes from the library at the Peabody Museum.*

DORCAS MACCLINTOCK

To MARGARET and PAMELA
who also like squirrels

INTRODUCTION

Squirrels belong to that largest and most varied of mammalian orders, the order Rodentia, which includes nearly one-third of all the land mammals and the great majority of the world's small mammals. Specialization among North American rodents ranges from the pocket gopher, which spends most of its life underground, to the volant flying squirrel. Rodent characteristics are large, strongly recurved incisor teeth, two upper and two lower, and a diastema (or gap) between incisor and molariform teeth. The incisor teeth grow throughout life from permanently open roots and usually appear chisel-like. Enamel covers anterior surfaces of the incisors and prevents the wear that erodes the softer dentine parts of the teeth.

Because the lower jaw of a rodent is loosely articulated the animal has a rotatory chewing motion. The masseter muscle, which extends from skull to lower jaw, has become variously modified in these gnawing mammals and provides the basis for the three rodent suborders: Sciuromorpha, Myomorpha and Hystricomorpha. Included in the sciuromorph group with the squirrels (family Sciuridae) are mountain beavers, pocket gophers, pocket mice, kangaroo rats and spiny mice, and beavers. In all these rodents the masseter muscle extends along the lower edge of the skull's zygomatic arch.

Members of the family Sciuridae occur in Eurasia and Africa as well as North and South America. Ground and tree squirrels are placed in the subfamily Sciurinae, while flying squirrels belong to the subfamily Petauristinae. All have the dental formula

$$\text{Incisors } \frac{1}{1} \text{ Canines } \frac{0}{0} \text{ Premolars } \frac{2\,(1)}{1} \text{ Molars } \frac{3}{3}$$

In North America squirrels are widely distributed and ecologically versatile; there are three main categories: *ground dwellers,* semi-fossorial and terrestrial squirrels; *tree squirrels,* highly specialized for arboreal life; *gliders,* the nocturnal flying squirrels.

Among the earliest known fossil sciurids several ground squirrels and a chipmunk have been recovered from Upper Miocene deposits, some twelve million years old; tree squirrels are known as far back in geologic time as Lower Miocene (approximately twenty-eight million years ago). These fossils indicate that the ground dwellers and the tree squirrels diverged early in the evolutionary history of the Sciuridae. By Upper Miocene the marmot line and the chipmunk line had diverged from the ground squirrel–prairie dog line. *Glaucomys,* the flying squirrel, has close relatives from the Pliocene of Europe and Asia but is known only from Pleistocene deposits in North America, where it is regarded as a relatively recent addition to the fauna.

On the evolution of squirrels in North America, and of their dentition (always the concern of the vertebrate paleontologist), Craig C. Black writes:

> The history of the Sciuridae in North America has been one of short evolutionary spurts and long periods of slow change. The early Miocene tree squirrels have changed remarkably little in their dentition over the last twenty million years. The spermophiles, after their origin in the late Oligocene, evolved very slowly through the Miocene and early Pliocene. The great diversification of ground squirrels is a relatively recent phenomenon that is still in progress . . .

Some North American sciurids are thought to have originated in North America and spread to Eurasia. The marmots and certain of the ground squirrels or spermophiles may have migrated west across the unforested Bering land bridge during the Pleistocene. Still other sciurid groups with North American origins have been restricted to this continent throughout their known history: *Cynomys,* the prairie dogs; *Ammospermophilus,* the antelope ground squirrels; *Tamiasciurus,* the red squirrel and its relatives.

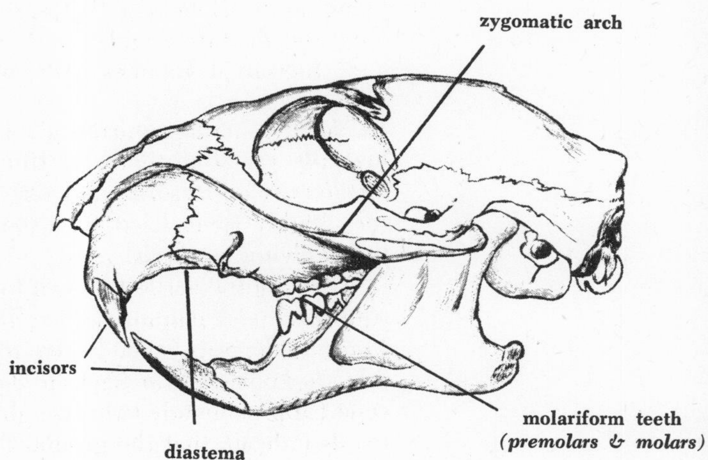

SQUIRREL SKULL
Prairie Dog (Cynomys)

I

THE GROUND DWELLERS

Marmots and Woodchucks, prairie dogs, ground squirrels and chipmunks are terrestrial squirrels. Primarily herbivores and seed-eaters, all the ground dwellers are equipped with internal cheek pouches. Some are virtually semi-fossorial in their habits; some live in colonies or "towns;" all dig their own burrows and for the most part have relatively strong digging claws.

Unfavorable times are avoided by the ability of many of the ground-dwelling squirrels to become dormant (aestivation and hibernation), sometimes for as long as eight months of the year. This group includes the true hibernating squirrels that through countless centuries have become adapted to climates with periods of extreme cold or of little or no moisture.

WOODCHUCK AND MARMOTS

Short-tailed, stout-legged, blunt-nosed and heavy-bodied, Woodchuck and marmots are the largest members of the squirrel family. In North America they are found in the mountains of Alaska and over most of Canada and the United States, except in the extreme south.

Marmota, the genus of Woodchuck and marmots, is derived from "marmotte," the common name of the European animal, *Marmota marmota,* in France. The expression *dormir comme une marmotte* is an idiom meaning, to sleep like a top, which is what all marmots do when they hibernate. The name "ground hog" obviously refers to the body shape, waddling way of going and ground-dwelling habits of Woodchuck and marmots. Everyone knows that on February second groundhogs emerge from their holes to inspect the weather. If they see their shadows, down they go into their burrows for six more weeks of winter. If the groundhogs cast no shadows, according to this bit of Woodchuck lore, spring is soon to follow.

In 1703 Baron La Hontan wrote a brief account of the Woodchuck in eastern Canada. *Siffleur,* or "whistler," was the name he gave them. The French voyageurs carried the name *siffleur* throughout the fur country of the northwest. An account by the eighteenth-century naturalist Mark Catesby in his *Natural History of Carolina* describes the Monax. This American Indian word meaning "digger" is still used for Woodchucks in parts of the southeast.

There are five species of marmots in North America. Instead of caching food as many tree squirrels do, marmots accumulate layers of fat, almost equal to half their body weight, and then hibernate for a considerable part of the year. Location and elevation of its habitat determine the length of time a marmot spends in dormancy.

The marmot skull is relatively large; its upper surface appears flat in profile. Triangular postorbital processes project at right angles to the length of skull and there is a conspicuous depression between the postorbital processes. Many rodents have orange incisor teeth, but those of marmots and Woodchucks are white. In all, marmots have twenty-two teeth—each side of the upper jaw having one incisor, two premolars and three molars, while each lower jaw bears one incisor, one premolar and three molars.

Woodchucks and marmots have five toes on each foot, although the thumb on the forefoot is so reduced that it does not show in tracks.

WOODCHUCK (*Marmota monax*)

Description. Almost everyone knows the Woodchuck. Even city dwellers have seen them feeding on fresh grass along the sides of highways. The Woodchuck waddles through the grass on short powerful legs. The broad head has a blunt nose and rounded ears which the 'chuck can close over its ear openings to keep out dirt. Woodchucks have bushy, somewhat flattened tails. Their black claws are strong and adapted for digging—four on each forefoot, plus a rudimentary thumb that bears a flat nail. Each hindfoot has five clawed toes. Woodchucks measure from sixteen to twenty-seven inches in length and weigh anywhere from four to fourteen pounds, depending upon the animal's age and season of the year.

The long coarse fur, sparse on the under parts, has a grizzled appearance. Generally the color is brownish with a grayish or reddish wash. The head is darker brown with buffy white on nose, chin and sides of face. Dark brown legs, black feet and tail hairs that are white-tipped complete the Woodchuck's coat color. There is much variation in Woodchuck appearance, and occasionally a pure white or nearly black Woodchuck is seen. Generally the young are not as richly colored as the adult Woodchucks. Molt occurs from May to September. Pelage (coat) change is observable first in the adult Woodchucks, with yearlings following suit followed by the young of the year. Molt commences with the tail, which appears thick and brightly colored in contrast to the worn and faded body fur. New hair appears on the nose and thick new hair slowly replaces the old frayed hairs over one part of the body at a time. Woodchuck young of the year do not always complete their pelage change. Cold may overtake them, sending them into hibernation only half-covered with new fur.

Distribution. With the exception of Florida and Louisiana, *Marmota monax* occurs throughout the eastern half of the United States. Its range also includes Canada, Labrador, and one subspecies, *Marmota monax ochracea*, is found in Alaska.

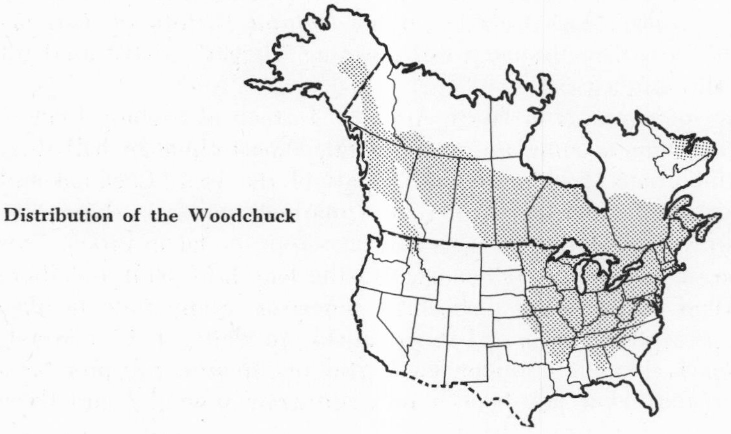

Distribution of the Woodchuck

Habits. Woodchucks are the least gregarious of the marmots. Usually they are solitary residents of burrow systems, although sometimes two, or even three or four, Woodchucks inhabit the same burrow. Preferred habitats include woodland edges, fence rows, meadows, stream banks and rocky bluffs. Once forest dwellers, Woodchucks have increased in numbers as woodland areas have been cleared for farming.

Woodchuck burrows, excavated beneath stone walls or among rocks, beneath tree roots or in meadows surrounded by abundant growth of grass and clover, are similar in their construction. A Woodchuck may use several burrows at different times, as well as maintaining both summer and winter residences. Woody or brushy areas are sites for the hibernating dens, while summer dens are located in meadows. The burrow's main entrance, marked by a mound of earth and stones, slopes at a moderate angle from the surface. Often it is excavated beneath a tree stump or rock. The dirt mound, which may be almost five feet in diameter, forms an observation post, sunning site and toilet, making it easy to determine whether or not the Woodchuck burrow is occupied. The musky odor secreted by the anal glands also indicates that a den belongs to a Woodchuck. Inside the main entrance, which is about one foot in diameter, the gallery slants downward to a depth of three or four feet. Depth of the burrow depends upon soil texture. If the soil is sandy a burrow may go as deep as six feet. Gravelly soil, with large stones, restricts burrow depth to four feet or less. At the bottom of the main entrance there is usually a large turn-around area. The burrow narrows gradually to about six inches in diameter. Inclining upward and to one side it may extend from ten to forty-five feet and lead to one or more enlarged nest chambers. The galleries that lead to the nests slope up and then down, to prevent flooding. The circular nest chambers measure no more than fifteen inches in diameter and are only half as high. Usually they contain nests of dry grasses or of dry grasses and leaves. The burrow may include blind tunnels and side branches. The Woodchuck deposits its long black scats in such side chambers or buries them in the entrance mound. As many as five side entrances, or plunge holes, make the burrow a safe refuge. These holes, well concealed by vegetation, appear to descend almost vertically for several feet and their four- or five-inch diameter is just large enough to admit a 'chuck in a hurry. Excavated by burrowing up from one of the main galleries, plunge holes do not have mounds of soil to mark their locations. Often Woodchucks, especially solitary individuals and old males, have a series of three burrows scattered across a field or along a fence line, and connected by a network of well-worn paths two to four inches wide. An acre may contain five or six such dens, which enable their occupants to wander over wide feeding territories.

To dig its burrow system the Woodchuck uses its strong chisel-like teeth to loosen small stones and cut through roots. Like most fossorial mammals, the Woodchuck has a heavy, relatively massive skull. Its hind legs push its body and with its strong chest and forepaws it shoves excavated soil out of the tunnel. Describing the speed of Woodchuck excavation, W. J. Hamilton, Jr., writes: "A woodchuck can bury itself from view in a minute, providing the soil is reasonably light and porous. Simple burrows with a single entrance, totaling five feet in length, are completed in a day." Dens are remodelled continually. Several times a week the burrows are cleaned out and fresh soil deposited at the entrance. Some 700 pounds of subsoil may be removed in the excavation of a single burrow system, contributing to aeration and mixing of the soil.

Woodchuck dens provide homes for skunks, raccoons, foxes and cottontail rabbits. Successive tenancy in a Woodchuck burrow in New York state was observed by W. J. Schoonmaker. When the original occupant hibernated, a cottontail rabbit found the 'chuck burrow a better winter shelter than its own above-ground "form." Subsequently a skunk moved in; then a raccoon. The last tenants were red foxes that turned the den into a family home. Other animals that sometimes use Woodchuck burrows for temporary refuge, snug winter retreats or as natal dens are opossums, weasels, otters, chipmunks, jumping mice, house mice, white-footed mice and short-tailed shrews. Quail, pheasant and ruffed grouse have been observed using abandoned Woodchuck dens, as have various kinds of snakes. When areas have burned over, Woodchuck burrows have saved the lives of various animals. Except for some use of their coarse-haired, thin-skinned pelts for fur and leatherwork and their parboiled flesh for food, Woodchucks themselves are of little economic value. But their construction of dens which serve as homes for animals that are important in controlling insect pests and rodent populations benefits the wildlife community as a whole. Because of

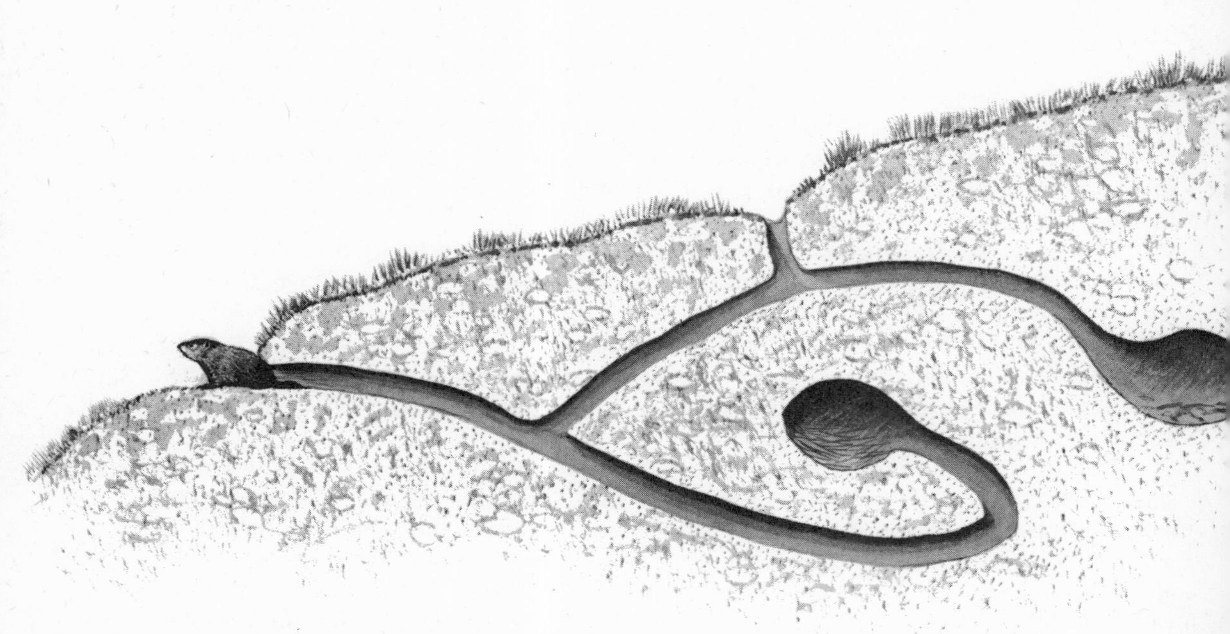

WOODCHUCK BURROW

their daytime habits Woodchucks are easily observed and provide enjoyment for anyone whose vegetable garden, cornfield or fruit trees are not in the Woodchuck's feeding territory.

Feeding and basking in the sun fill the few hours each day that a Woodchuck is out of its den. Always alert, a Woodchuck freezes momentarily, then dashes for its burrow when danger threatens. The Woodchuck's preferred gait is an ambling walk, but when necessary it can gallop heavily up to a speed of ten miles an hour. Like bears, raccoons, and skunks, Woodchucks are plantigrade, or flat-footed.

Before leaving its den a Woodchuck pauses about three feet inside the doorway and listens. It may give a low-pitched whistle, followed by a tremulous call or warble. This habit has earned the Woodchuck the name "whistle-pig." Cautiously the Woodchuck's head appears. Highly set eyes, ears and nose enable the Woodchuck to see, hear and smell while only the top of its head protrudes above the burrow's rim. The Woodchuck climbs out on the entrance mound and slowly waddles down a trail to a feeding place, pausing every few minutes to stand up on its hind legs and survey the landscape. As the Woodchuck feeds, sitting up on its haunches, its nimble forepaws are used as "hands" to grasp grass stems.

Early morning and late afternoon to evening are times of greatest Woodchuck activity out of the burrow. By mid-summer 'chucks prefer the early evening feeding time. Sun-basking takes place atop the earthen entrance mound or on a nearby rock. The rest of a Woodchuck's day, as much as twenty three hours, is spent below ground. Its fossorial life is not well known, but probably only part of its time underground is spent sleeping.

Woodchucks rarely stray more than a hundred yards from their home burrows. Size of home range depends upon den location. If the burrow is in a gully or woodlot, its occupant travels several hundred yards to feeding areas. Woodchucks that den in clover or alfalfa fields restrict their home ranges to a twenty-yard radius of the burrow. Commonly a Woodchuck's home range measures one hundred yards to an eighth of a mile in diameter.

Where 'chucks are abundant their feeding territories overlap, but home dens are staunchly defended. The Woodchuck that bounds down a neighbor's burrow in alarm is promptly run out again. A study of Woodchuck populations in the middle Atlantic states estimated twelve to eighteen animals per one hundred acres. In New York state, where Woodchuck populations are sometimes as dense as five per acre, the average number of Woodchucks is about one to every three acres.

Woodchuck diet includes leaves, flowers and soft stems of various grasses, field crops such as red and white clover, alfalfa, chickweed and many wild herbs. As much as a pound of green food is consumed each day by a Woodchuck. Peas, beans and corn, melons and apples are favorite foods. Sometimes Woodchucks damage trees by eating bark. Rarely carnivorous, Woodchucks may eat an occasional grasshopper or egg or young of a ground-nesting bird.

Much of a Woodchuck's life is spent in hibernation. Depending upon climate, four, five or even more months are spent in deep slumber. As autumn days shorten and grow colder Woodchucks become noticeably fatter and are less active. In early July Woodchucks begin to take on weight. A layer of fat is laid down first in the inguinal region. This spreads over the hindquarters as summer progresses and finally is joined by a fat layer over the shoulders to form a blanket of fat that is nearly a half

WOODCHUCK
(*Marmota monax*)

inch thick. Accumulation of fat, causing lethargy, is believed to be a motivating factor in hibernation. During late September Woodchucks become scarce. Those that live in the Adirondacks of New York state are curled up asleep in underground nests by this time. C. H. Merriam describes the change in activity:

> In summer, throughout the farming districts, they commonly leave their burrows early in the morning, late in the afternoon and during moonlight nights, but may sometimes be found abroad at all hours. As autumn approaches, and they become more and more fat and sleepy, they usually appear only in fine weather, and then but for a few hours in the hottest part of the afternoon.

Hamilton observed several Woodchucks on a late September afternoon, stretched out "as though dead" at their burrow entrances. The next afternoon he saw no 'chucks.

Older, fatter adults are the first to succumb to sleepiness. Followed by the leaner adults and young of the year the Woodchucks move to their winter dens in woods and hedgerows. Sometimes two Woodchucks hibernate in the same den. Before going to sleep for the winter the Woodchuck seals itself in by plugging off the hibernating chamber with earth from the main burrow tunnel. Rolling itself into a ball, the 'chuck rests its head between its hind legs and draws its forepaws underneath its body. The hibernating chamber is usually well protected, sometimes beneath the tangled root system of a tree, making it almost impossible for a predator to reach the somnolent animal. Woodchucks have also been known to burrow deep into hay stacks for their winter sleep.

By late October Woodchucks everywhere have gone into hibernation. In this torpid state the Woodchuck's body temperature falls to between 38 and 57° F. Breathing is very slow, one breath every six minutes, and heartbeat and circulation are reduced in rate. Waking a Woodchuck by warming it up requires several hours. During mid-winter mild spells Woodchucks sometimes leave their burrows. A drop in temperature causes them to resume their winter sleep.

In late February or March, depending upon latitude, Woodchucks emerge from hibernation. Again it is the largest that are first to awaken. The 'chucks are lean, the stored body fat is used up. Weight loss averages nearly 40 percent of body weight during hibernation. This loss continues for several weeks after emergence because little green food is available. Bark of small trees supplies some nourishment, and the 'chucks leave ragged tooth and claw marks on trees near their dens.

In early spring Woodchucks are active above ground at any time of day or night, and they wander widely. The males begin their treks in search of mates. Woodchuck tracks, often observed on spring snow, are four-toed in front, five-toed behind. When galloping, fore and hind feet are closely bunched, the sets from twelve to twenty inches apart. About four inches separates the tracks made by a walking Woodchuck. Squeals, snapping of incisor teeth and grinding of cheek teeth mark the battles of Woodchucks that chance to meet at this time of year. The fight may be over territory or a prospective mate. Woodchucks sometimes lose pieces of ears or parts of their tails in combat. Spring also is moving time for Woodchucks. Their winter dens are vacated and they move to their summer or field burrows.

Most male Woodchucks mate with several females each spring, but sometimes a male will stay with his mate after birth of the young, even helping with the care of their offspring. Baby Woodchucks are born anytime from late March to mid-May, after a gestation period of thirty-one to thirty-three days. Litter size ranges from two to nine, with the average containing four dark-pink, wrinkled babies. At first the young are blind, less than four inches long and about an ounce in weight. Short hairs sprout over their eyes and on their cheeks. When nursing, the mother Woodchuck either stands over her young or sits up, bear-fashion. At one week the babies' backs have become heavily pigmented, and their snouts and foreheads covered with short grayish fur. By the time the babies are four weeks old, their eyes open and they clamber about in the burrow. Mother Woodchuck brings green food into the burrow for her young, for the babies seldom venture far outside the burrow until their sixth or seventh week. By this time they are grizzled grayish brown, with black fur covering the top of the head. The little (about a pound and a quarter) Woodchucks often wrestle playfully. In mid-summer their lives take a serious turn when the mother 'chuck drives them off to live by themselves in burrows they dig nearby. In fall the young Woodchucks wander off to find their own territories even though they do not become mature until their second year. This yearly shifting of Woodchuck families, as well as the wide-ranging springtime quests of males, lessens the chances of inbreeding.

Except for irate farmers and gardeners the ever-alert Woodchuck has few enemies. Dogs and foxes are its most dreaded foes. Foxes take only young Woodchucks, as an adult is more than a match for a fox or smaller dog. Badgers, coyotes, bobcats, as well as large hawks and owls and large snakes occasionally snatch young Woodchucks. Bears sometimes dig Woodchucks out of their burrows. Such parasites as worms, ticks, fleas and warbles, or fly larvae, plague Woodchuck populations. Five or six years is the natural Woodchuck life span, although individuals have been known to live for ten years.

Woodchucks are good swimmers and there are numerous accounts of their aquatic habits. W. E. Cram observed a swimming Woodchuck:

> One summer morning . . . as I approached the bend of a woodland stream, I saw a woodchuck slip through the undergrowth and plunge down the bank of the stream into the water. He dived like a muskrat, and in the clear water I could see him a foot or more under for several yards before he came to the surface. He then swam for several rods in a half circle with head and shoulders out, and, landing on the same bank, disappeared among the bushes.

Another surprising Woodchuck habit is tree-climbing. When pursued, Woodchucks have been known to climb as high as fifty feet. Numerous observers have reported 'chucks taking to trees and grasping branches eight to ten feet above ground to elude barking dogs below. Woodchucks sometimes are seen atop wooden fence posts, sunning themselves or surveying their territory, or walking along fence rails.

Like all rodents, marmots have incisor teeth that grow throughout life, except during hibernation. In chewing, the lower jaw moves forward and backward and the incisors occlude. This wear keeps them in shape, although Woodchuck skulls have been found with incisor teeth that have failed to meet. Hamilton estimates malocclusion in one percent of Woodchuck populations. Growing persistently, the malformed incisors curve out of the mouth in various fashions. In such instances of malocclusion the sharp tips of the growing incisors may curl upward to pierce the hard palate, an eye or even the Woodchuck's braincase, causing death. Or they may coil about the animal's jaw, locking the jaw and causing death by starvation. Sometimes the spiral growth of incisors gives the Woodchuck a formidable appearance, but does not interfere with its life.

A spectacular example of malocclusion and the unrestricted development of upper and lower incisors is found in a Woodchuck skull (number 945, Peabody Museum of Natural History, Yale University), described by M. R. Thorpe in the *Journal of Mammalogy*. The upper incisors curve to the left and their one and one-quarter coils measure four and three-fourths inches. The right upper incisor has penetrated the left maxilla and grown down into the mouth again through the left incisor tooth's opening. The lower incisors, growing to the right, measure three and a half inches and have worn away the right nasal bone. Although this Woodchuck obviously had difficulty opening its mouth, it was a fat and healthy animal when collected near Woodsfield, Ohio.

YELLOW-BELLIED MARMOT (*Marmota flaviventris*)

Description. Yellow brown undersides give this marmot both its common and species names. Sometimes it has been called the Yellow-footed Marmot. Grizzled brown fur covers the upper parts of its body. On top of its head the fur is black and a whitish band crosses its nose. Total length of the Yellow-bellied Marmot ranges from nineteen to twenty-eight inches, of which the tail measures five to nine inches. Weight varies from four to twelve pounds, depending upon season and age of the marmot. Occasionally, Yellow-bellied Marmots weigh as much as seventeen pounds. Because its burrows are almost always located in rock slides or beneath a jumble of boulders, this marmot is known, especially in the West, as the "rockchuck."

Distribution. Yellow-bellied Marmots occur in south-central British Columbia. Their range extends south through the Rocky Mountains to New Mexico and south through the Cascades and mountains of eastern Washington and Oregon to the southern Sierra Nevada. They also are found in the higher ranges of central Nevada. Excavations of kitchen middens in northern Arizona as well as bones found in a cave in California's Providence Mountains indicate that the range of the Yellow-bellied Marmot once extended farther south than it does today.

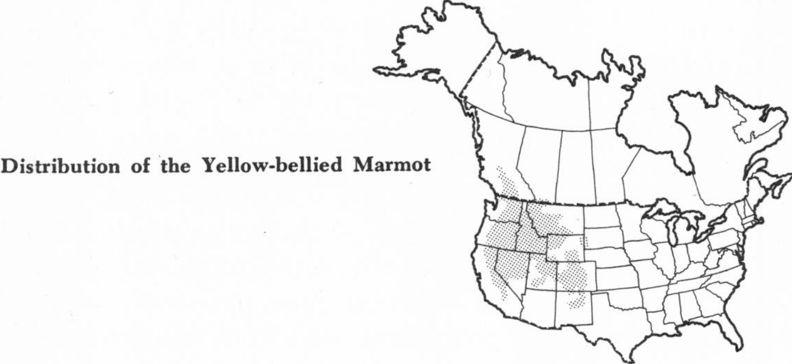

Distribution of the Yellow-bellied Marmot

THE WAYS OF A YELLOW-BELLIED MARMOT

An osprey wheeled over the curve of the Snake River, casting its shadow over rock-strewn areas of sagebrush and grasses. In the distance the snow-flecked Tetons jutted above the floor of Wyoming's Jackson Hole valley, forming a spectacular granite skyline. Along the river terrace below the osprey could see several marmots—part of a colony that included some thirty-five Yellow-bellied Marmots and covered about seventeen acres. Although the marmots paused in their feeding to survey their surroundings, they seemed not to be alarmed by the osprey gliding overhead. In its flights along the river the osprey often sighted scattered colonies of rockchucks.

On this July day the colony was an active place. The marmots emerged from their burrows as the sun's rays first reached the colony. For more than fifteen minutes each rockchuck sat grooming and sunning itself, before seeking its breakfast. Nearby, on the river's floodplain, was a lush growth of timothy and white clover, and by eight o'clock that morning almost all the marmots in the colony were feeding there. Their crawlways wound and criss-crossed through the grass. As they fed, the marmots lay down and crawled along, now and then raising their heads while they chewed. Marmots feeding near the edges of the timothy-clover patch were more alert and frequently sat up on their haunches. These were the colony's sentinels, whose warning whistles would cause the marmots to dash for their burrows.

The burrow systems of Yellow-bellied Marmots consist of two types of dens. The home burrows are used by the rockchucks for sleeping quarters and as nurseries. When alarmed, each marmot makes every effort to return to its own home-burrow, to plunge to safety through one of its three or more entrances. Each marmot also has auxiliary burrows, used for refuge when the home burrow cannot be reached. One or two entrances lead into these auxiliary burrows, which are connected by a network of trails leading to various feeding grounds.

Along such a marmot-made trail ambled a young rockchuck, his tail held high and waving jauntily from side to side. He had been born late in May, one of a litter of three. Like some of the adults in the colony he was very dark brown—almost black—and lacked the yellow belly of his species. Melanism, the overabundance of dark pigment, melanin, in skin and hair, occurs often in the marmot colonies in the Jackson Hole country. The young marmot still denned with his mother and two sisters. Their home burrow was located near the center of the colony, its entrance beneath a large boulder. Burrows along the periphery of the colony belonged to last year's young rockchucks. The young marmot's father also occupied a home burrow near the middle of the colony. The only adult male in the colony, his harem included thirty-one female marmots. After emerging from hibernation in late March, most of the adult female marmots in the colony had mated with the large male. Now the year's young were almost two months old and somewhat larger than ground squirrels. They were playful, romping and tumbling with their litter-mates. Sometimes a mother marmot played with her babies outside the burrow entrance.

Another young marmot appeared on the trail, and the two animals exchanged greetings—cheek-sniffing and tail-arching. Older marmots also arched tails and bodies when they met, but often a chase ensued, one marmot shrieking angrily at the other's heels. Near its home burrow each marmot appeared self-assured and signalled its aggression to passing marmots by tail-flagging and grooming. Submissive marmots indicated their feelings by slinking along with tails down and by allowing themselves

YELLOW-BELLIED MARMOT
(*Marmota flaviventris*)

to be groomed. Some of the marmots in the colony occupied separate home ranges. Others had ranges that overlapped, although seldom were two marmots active in the same area of overlapping territory at the same time. Angry chasing was often directed at the year-old marmots, whose burrows rimmed the colony, causing them to leave the colony and dig burrows elsewhere. Such behavior tended to keep the population of the marmot colony from becoming too high.

A sudden high, sharp alarm call pierced the air. The marmots stopped feeding and galloped back along the trails toward their burrows. In fright, their tails streamed behind them. Two ravens, feeding on an elk carcass across the river, were the cause of the alarm call. During summer the appearance of moose, deer, bears, coyotes, horses and fishermen at times spread alarm in the marmot colony. Sometimes only a shrill whistle was sounded—an alert when danger was not imminent. Then the marmots responded by sitting up and looking around. More often an intruder caused the higher, sharper alarm call to sound, and the marmots scurried for the safety of their burrows. Sight of a sandhill crane, great blue heron or marsh hawk near the colony was regarded as a threat by the marmots. But none of these predators took a toll of more than a few young marmots each season.

After the morning feeding the rest of the marmots' day was spent sunning, digging, or resting in their burrows, with occasional forays for food. Sunning in the morning, the marmots lay sideways toward the sun, enjoying the warmth of its rays. As the day became hotter the marmots oriented themselves lengthwise to the sun. By mid-day the sunning marmots sought shade of boulders and log piles. In cloudy or stormy weather the marmots preferred to stay in their burrows.

Here and there in the colony a marmot renovated its burrow, digging with its forelegs and throwing the dirt between its hind legs. Then, lying down and shoving the loose dirt with its chest and forelegs, the marmot removed the clutter from its burrow entrance. Sometimes a marmot would lie down, tear some grass, use its forepaws to fill its mouth and carry the grass to its burrow for nest lining.

Late afternoon brought a second period of activity to the marmot colony. After eating their fill the marmots returned to their dens, lingering near the burrow entrances during the last hour of light. By a half hour after sunset all the marmots had disappeared into their burrows for the night.

In late summer decreased day length and colder early-morning temperatures caused changes in the marmots' way of life. They emerged from their burrows later, fed more avidly and went underground earlier.

During late August and September activity waned in the marmot colony, as the animals went into hibernation. The young melanistic marmot would spend his first long sleep in his mother's burrow, rolled into a ball, his forepaws over his eyes.

HOARY MARMOT (*Marmota caligata*)

Description. This is the marmot that wears boots as its specific name *caligata* implies. Largest of the American marmots, the Hoary Marmot appears mostly black and white, with blackish feet. Actually the fur is gray-black, tipped with white. Buffy red tinges the fur of the hindquarters; the underfur is soft and dense; the tail, longer than that of other marmot species, is dark reddish brown. Conspicuous black streaks mark each side of the head and neck. A transverse black band on the snout separates white nose patches from white patches in front of the eyes. Parkas and robes are made of Hoary Marmot pelts by Indians and Eskimos. Hoary Marmots range from twenty-five to thirty-one inches in length and weigh five to fifteen or more pounds.

Distribution. Hoary Marmots are found in the mountains from northwestern Alaska and the Alaska Peninsula southward to central Washington and Idaho. Except in Alaska, where their burrows are dug on open grassy hillsides or out on the flats, Hoary Marmots live near or even above timber line, always in or about talus slopes near grassy alpine meadows.

Distribution of the Hoary Marmot

Habits. Whistlers *par excellence,* Hoary Marmots are always on the alert and emit loud, piercing whistles when excited. Reportedly their shrill alarm whistle carries for more than a mile. The repertoire of a Hoary Marmot also includes a range of barks, yips and yells.

Most gregarious of the marmots, Hoary Marmots live in villages in their talus habitats. Saxifrage and stonecrop grow in small rock crannies, and small patches of grass flourish among the rock rubble—food for the Hoary Marmot. Berries and roots are also eaten. South-facing slopes generally support more life, both plant and animal, than the north-facing sides of valleys. The Hoary Marmot, fond of sunning on boulders, prefers the daytime warmth of south-facing slopes. Sharing its rocky environment are chipmunks, pikas and woodrats. Sometimes a porcupine wanders through. Several groups of Hoary Marmots may inhabit an extensive rock slide area. Each group has its sentinel, usually an older female, surveying from a boulder lookout. A golden eagle soaring overhead evokes a warning whistle that echoes from surrounding cliffs and crags and causes mountain sheep, grazing on the high slopes, to raise their heads and test the wind. In addition to eagles, the Hoary Marmot's enemies include hawks, wolves, coyotes, cougars, lynx and bobcat, and grizzly bears. Burrows beneath the boulders usually provide safe refuge, though persistent grizzly bears sometimes succeed in digging out a Hoary Marmot burrow. The only description of a Hoary Marmot home is given in *The Grizzly Bear* by W. H. Wright:

HOARY MARMOT
(*Marmota caligata*)

The den ran in under several layers of loose flat rocks, some of which were two or three feet long by half as many wide, and several inches thick. These he had ripped out easily and thrown down hill, and the dirt and small boulders had been hurled out and now covered the snow all about for a space of ten or twelve feet.

On the rocks and snow were large spots and blotches of blood, telling of the feast that had rewarded his labors, and that there had been more than one marmot was shown by the numerous tracks. These animals had burrowed down some six or seven feet into the side of the mountain, and under a large flat stone they had scooped out a little cave, some three feet in diameter, where they had a soft bed of grasses that they had carried in. When the grizzly broke his way into their home there had been a great rush for freedom.

By late September or early October Hoary Marmots retreat into their burrows to sleep through the long cold months. When they emerge, sometime in April, new plant growth will feed them. Hoary Marmots living in the Arctic hibernate from September to May. Even at so late a date they may have to tunnel through snow to emerge from their dens.

In true marmot fashion the first order of business for males awakening from a winter's sleep is finding a mate. A month later two to five young Hoary Marmots are born. The youngsters remain with mother marmot during the summer and may even spend their first winter in her den.

OLYMPIC MOUNTAIN MARMOT (*Marmota olympus*)

Description. This marmot resembles the Hoary Marmot. Its color is drab brownish mixed with white. During summer its coat color bleaches to yellow above. It has a whitish nose, a transverse nose band and white patches in front of its eyes. But nose band and patches are not as distinct as those of the handsomer Hoary Marmot.

Distribution. At home in the alpine meadows of the Olympic Mountains of Washington, this marmot burrows under rocks and on open slopes. A colony of twelve to fifteen Olympic Mountain Marmots was observed to feed on sedges, lupines, lilies and some Composites, as well as miner's lettuce, heather blossoms and mosses. This colony had its sentinel, an old marmot whose burrow was highest on the ridge.

VANCOUVER ISLAND MARMOT (*Marmota vancouverensis*)

Description. Except for its uniformly dark brown color the Vancouver Island Marmot looks much like the Hoary Marmot. One mammalogist has suggested that the Hoary, Olympic Mountain and Vancouver Island marmots are subspecies of the European marmot, *Marmota marmota.*

Distribution. This marmot occurs in the mountains of Vancouver Island, British Columbia, only a few miles across the San Juan Strait from Washington.

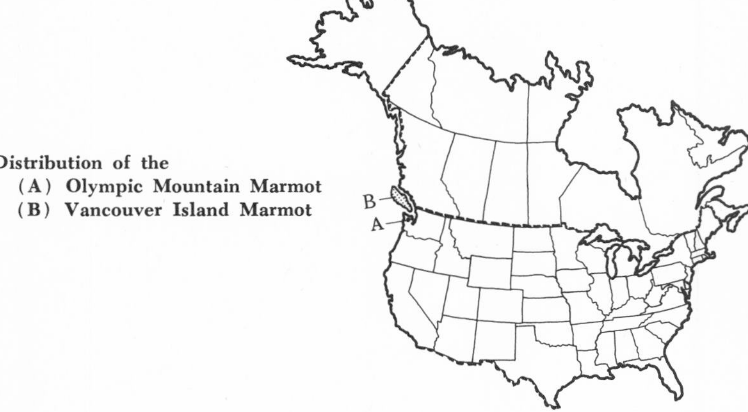

Distribution of the
(A) Olympic Mountain Marmot
(B) Vancouver Island Marmot

PRAIRIE DOGS

> The prairie dog is, in fact, one of the curiosities of the Far West, about which travellers delight to tell marvelous tales, endowing him at times with something of the politic and social habits of a rational being, and giving him systems of civil government and domestic economy, almost equal to what they used to bestow upon the beaver.

So writes Washington Irving in *A Tour on the Prairies.* After visiting a thirty-acre prairie dog village on his 1832 expedition Irving records that prairie dogs were "quick, sensitive, and somewhat petulant. . . . At sight of us, the picket guards scampered in and gave the alarm; whereupon every inhabitant gave a short yelp, or bark, and dived into his hole, his heels twinkling in the air as if he had thrown a somerset."

These stout, short-tailed, short-legged burrowing squirrels, smaller than marmots but larger and stockier than ground squirrels, belong to a genus that is strictly North American. First collected by Lewis and Clark, they were described by George Ord as a species of marmot and given the scientific name *Arctomys ludovicianus,* the marmot of Louisiana Territory. Lewis and Clark entered brief notes on prairie dog behavior in their journals. In 1817 the naturalist Rafinesque proposed the generic name *Cynomys,* or dog-like mouse, suggested by the prairie dog's distinctive barking call. Common names of the prairie dog include barking squirrel, prairie squirrel, petit chien, prairiehunde and wishtonwish.

Prairie dogs measure twelve to fifteen and a half inches in length and weigh anywhere from one and a half to three pounds. Their coarse, close-lying fur is grizzled yellow-gray or pale buffy above and grades into white on the underparts. Buffy white fur forms eye rings and covers cheeks and sides of nose. Prairie dogs are divided into two groups, the Black-tailed Prairie Dogs (subgenus *Cynomys*) and the White-tailed Prairie Dogs (subgenus *Leucocrossuromys*).

The two subgenera of prairie dogs exhibit different ways of life that are correlated with habitat variations. Living for the most part in mountainous areas with broken terrain, prairie dogs of the white-tailed group have habits similar to rock squirrels and live in pairs or widely separated family groups. Black-tailed Prairie Dogs long ago congregated in large numbers where grasslands afford easily excavated substrates. Digging their burrows in common areas made it possible for the colonies to modify the grassland vegetation, maintaining a stage favored by the prairie dogs. Social structure developed in the towns. Like all rodents, prairie dogs have acute senses, especially of hearing and smell. Their safety, which lies in fleeing to their burrows, is further insured by their habit of giving alarm signals.

BLACK-TAILED PRAIRIE DOG (*Cynomys ludovicianus*)

Description. A comparatively long tail, more than one-fifth the animal's total length, tipped with black or brown, distinguishes this species, and the Mexican Prairie Dog (*Cynomys mexicanus*), from prairie dogs of the white-tailed group. Black-tailed Prairie Dogs are somewhat larger than their white-tailed cousins, and

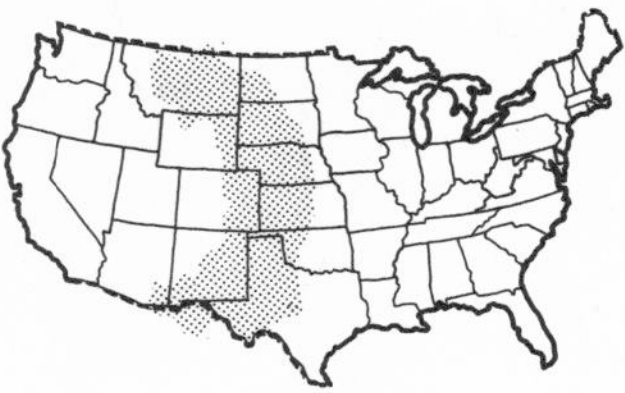

Distribution of the Black-tailed Prairie Dog

their more massive skulls bear large cheek teeth that are noticeably expanded laterally.

Distribution. Black-tailed Prairie Dogs originally occupied "towns" that extended over vast areas of the short-grass plains from southern Saskatchewan and central North Dakota south to southeastern Arizona and Texas. Their survivors now live in small colonies protected in national parks and wildlife refuges.

A hundred years ago, before the arrival of the white man, North America's Great plains supported a variety of living things, all to some extent dependent on one another. Enormous herds of buffalo in numbers that "darkened the whole plains" traversed these grasslands. Bison and prairie dog exhibited a mutual interaction that favored both. Heavy grazing by bison changed the grassland succession. Tall and medium height grasses, such as bluestems and needlegrasses that form the climax vegetation, gave way to subclimax short grasses dominated by buffalo grass and grama grass. Prairie dogs prefer heavily grazed areas for their townsites. Once the short grasses took over prairie dogs maintained the subclimax. This in turn benefited the bison, which prefers short grasses.

To the prairie dog's consternation the bison also favored its towns as bare dusty areas for wallowing. Itching caused by sharp-pronged grass seeds, flies and mosquitoes made dust wallowing the buffalo's delight. Over the big animal rolled, squirming and kicking, until part of the prairie dog village was flattened into a saucer-shaped depression ten or so feet in diameter—a buffalo wallow.

The Plains Indians were part of the balance that nature had evolved. They roamed the prairie for hundreds of years, their temporary villages having little more effect on the plains than did the prairie dog towns. But the white man was not part of this plan. Seeing wealth in the grasslands in form of cattle and sheep he exterminated the competing bison and pronghorn, poisoned prairie wolves and coyotes and prairie dogs, eventually turning large stretches of the plains into deserts.

THE WAYS OF A BLACK-TAILED PRAIRIE DOG

As the sun rose, the crater-like mounds of a small Black-tailed Prairie Dog town cast shadows over the short-grass plains. A remnant of the towns that once extended for hundreds of square miles, this prairie dog colony survived because a wildlife refuge, some 1,200 acres of short-grass prairie, had been formed in the eastern foothills of the Rocky Mountains in Wyoming. One by one the prairie dogs climbed out on their burrow mounds to begin the day's activities. Buffalo grass, wheat grass and grama grasses waved gently on the high plains' dips and swells, dotted yellow by sunflowers, butterweed and broomweed.

Foot-high mounds, ten to thirty feet apart, marked the burrow entrances of the prairie dog town. Bare ground surrounded each crater. The heavily worked vegetation appeared crisscrossed by trails. Atop one mound sat a female Black-tailed Prairie Dog. Moments before she had left the nest chamber of her den and clambered up the steep plunge hole to bask in the early morning sunshine. It was mid-June, and like most of the female prairie dogs this one had youngsters. Her babies had been born in April after a month-long gestation period. Some of the litters were as large as ten, but most of the prairie dogs had produced four, five or six babies. A few of the younger females had only two or three offspring. The young prairie dogs' eyes opened when they were five weeks old and they began to crawl about the burrow. By the seventh week they had learned to follow their mother out onto the mound. Curious but shy, the young prairie dogs pressed together, craning their necks to look about the colony. The mother prairie dog sat upright, allowing her four "pups" to nurse briefly. The weaning process had begun as green plants they nibbled in their first days above ground supplemented their diet. As soon as the young are weaned mother prairie dogs leave their burrows to their offspring and move into empty burrows or dig new ones. In the next two and a half months the young prairie dogs would almost double in weight. Their soft, fine, pale-colored fur soon would be replaced by the coarse fur of the older prairie dogs. Then, one by one, the young prairie dogs leave the home burrow and seek new homes nearby.

Each prairie dog town is divided into *coteries, clans* or *precincts*. Whatever term is applied, these are the most important subdivisions of the community, assuring protection against predators and regulating prairie dog populations. Coteries are the basic units of prairie dog social organization. Each covers approximately seven-tenths of an acre. The members of a coterie use the various burrows within the territory freely and spend considerable time grooming each other and "kissing" in the manner of prairie dogs. Seeing another of its species, a prairie dog will run up and rub noses, thus identifying an animal that belongs in the same territorial unit. Coterie members stake their boundaries. At intervals they jerk their heads up and back, with forefeet in air, and utter their characteristic bark. Then they drop to all fours again. Thus challenged, a prairie dog that belongs to another clan scurries off. Territorial yips draw replies from neighboring mounds, each burrow occupant atop its mound jerking into the air to give its call. This territory-staking has been likened to the performance of male sage grouse on their booming grounds and to the singing of such birds as horned larks and longspurs over their grassland territories.

In spite of the antagonism that exists between prairie dogs of different coteries, all prairie dogs in a town cooperate in watching for any suspicious movements overhead or on the ground. Such threats of danger cause a high-pitched yip that is a warning call, a signal alerting prairie dogs throughout the town. Such a sudden shrill alarm call caused the mother prairie dog to sit up and listen. Again the call came, this time higher-pitched and more rapid. Prairie dogs dashed for safety. The young prairie dogs raced after their mother and plummeted into their burrow. The female stayed on the mound, joining some of the other adult prairie dogs in a chorus of yips and barks of fury and alarm. Then dropping to all fours, she gave a flicker of her tail and dove into her burrow. This time the alarm had been caused by a golden eagle, swooping on set pinions and flashing over the town as the startled prairie dogs fled for their burrows. One prairie dog was snatched aloft to become a meal for the eagle as it perched on a nearby rock outcrop.

Curiosity is a prairie dog trait, and even as the eagle threatened the town many of the prairie dogs had paused a short distance inside their burrows. Soon they began to reappear, and their slightly rounded heads, with eyes set high, enabled them to see over the mound rims without exposing too much prairie dog. Each burrow entrance, six to eight inches in diameter, narrows abruptly to an almost vertical plunge shaft four or five inches wide. The plunge shaft extends three to sixteen feet below ground. Where the short entrance slope gives way to the plunge shaft each burrow has a vestibule, guard room or listening post—a shelf or niche dug into the side of the burrow. This is where many of the town's inhabitants had paused to listen in their burrows when the eagle swooped overhead. Deep in each burrow a round, eight to ten inch in diameter nest chamber is lined with mats of grass that the prairie dog occupants bundle between their jaws and carry into their nests. On the next page, W. H. Osgood gives an account of a prairie dog burrow he excavated:

BLACK-TAILED PRAIRIE DOG AND ITS BURROW

> . . . burrow went down vertically to a depth of 14½′ below the surface, when it turned abruptly and became horizontal . . . horizontal part was 13½′ in length. One-third of the horizontal part, the terminal four feet, and two old nests and passageways were plugged with black earth brought in from the surface layer, which was very different from the light-colored clayey earth in which the greater part of the burrow lay. . . . Four or five feet below the entrance was a diverticulum, or short side passage, probably used as a place in which to turn around when the animals come back to take a look at the intruder before finally disappearing in the bottoms of their burrows. It is used also, apparently, as a resting place where they bark and scold after retreating from the mouths of the burrows . . .

The five (rather than four, as in most squirrels) sharp, rather long claws on each front foot are the prairie dog's digging tools. Loosened earth is kicked out of the hole by backward strokes of the hind feet. When a prairie dog is too far below the surface to kick dirt all the way up and out of the shaft, it comes out of the burrow and again begins at the top to kick soil out of the hole. Where the plunge hole ends and the den turns horizontally and upward, the burrow may fork into side passages or loop hallways. Usually the den has only one entrance. But some prairie dogs push an escape tunnel close to the ground surface, as a hatchway through which they can dig free when pursued underground by badger or ferret, or when flooding threatens their tunnels. Burrows abandoned or left in disrepair by prairie dogs are sometimes taken over by Richardson's Ground Squirrels.

The mother prairie dog emerged again, this time responding to the all-clear calls that sounded all over the town. Her youngsters scampered over to a nearby burrow where other young prairie dogs were playing. The female began to work on her mound—important work for any prairie dog. The dike-like mounds prevent flooding—water from summer cloudbursts sometimes stands two or three inches deep over the flat, hard-baked prairie soil. Especially after rains prairie dogs work to repair their burrow cones. Some prairie dogs construct mounds as high as two feet and as much as four feet in diameter. With her paws the prairie dog heaped up soil, tamping it into place with repeated drives of her nose. Now and then she stopped to clean her dirt-clogged digging paws. The curved position of her body drove powerful thrusts of her nose against the soil mound, leaving small nose prints over the inside of the crater. Often the adobe-like mounds are reinforced with dried grass stems, which are worked in with the soil. When floods threaten a town, poorly tended dikes often give way. Water seeps down the tunnels and the inhabitant scrambles out to be marooned on its mound. When prairie dogs are flooded out, they usually swim to higher ground where new homes are dug immediately and a new prairie dog town formed.

Indians are said to have drowned out prairie dogs for food, and there is at least one report of a clever coyote that built a dam to divert a flash flood into a burrow and minutes later seized the fleeing occupant. Others attest that many gallons of water are required to drown out a prairie dog. This may be because the prairie dog has perched high and dry in its guard room or is snug in an air pocket that has formed in an elevated chamber of its burrow system.

The prairie dog mother and four youngsters sat erect on their mound, looking about. Dropping to all fours, they wandered away from the burrow to feed. Grasses and forbs form the bulk of prairie dog diet, and prairie dogs are such gluttons that more than half their above-ground time is devoted to feeding. Stems, roots and seeds are devoured. A stem or leaf is grasped in one paw and fed into the rapidly working jaws. Insects, mostly cutworms and beetles, are also eaten. Grasshopper swarms that once passed over and alighted in prairie dog towns provided feasts. L. C. Whitehead witnessed such an event in 1923 when a Texas prairie dog population was out in full force in pursuit of grasshoppers. Some of the prairie dogs jumped to catch the insects, and gave chirping barks. For drink prairie dogs have only puddles that form after infrequent showers, and so are dependent upon their ability to produce metabolic water.

From her burrow mound the mother prairie dog watched a box turtle pry over a buffalo chip in search of dung beetles. The same turtle, or tortoise, had buried itself in a side tunnel of her burrow and slept away the preceding winter. Sometimes toads also are peaceful inhabitants of prairie dog burrows.

Hot, dry winds of summer began to parch the prairie soil, withering and yellowing the grass. Under the baked soil surface grass roots stretched down deeper and deeper for moisture. Heat waves shimmered under the blazing sun. The prairie dogs took long mid-day siestas, emerging from their burrows in the morning cool and again near sundown.

Two burrowing owls perched on the rim of a distant mound. As their large yellow eyes scanned the colony the little owls bobbed and curtsied incessantly. The prairie dog town was home to twenty or so of these small, long-legged owls that nested underground in chambers of abandoned prairie dog burrows. Young burrowing owls, like prairie dog babies, were also out in the sunshine, taking their meals on their earthen doorsteps and making their first attempts to fly by hopping and frantically beating their wings.

The presence of burrowing owls and prairie rattlesnakes in prairie dog towns gave rise to many cowboy tales about prairie dogs, owls and rattlesnakes happily living together. Charles M. Russell, the Western artist, offers this whimsical account:

> It seems that long ago there came a time when there were less and less prairie dogs. This was because eagles and hawks and even big owls found they were quicker than fat prairie dogs. While Mr. Prairie Dog and his family were frisking around never looking up for danger, these big birds would swoop down and carry off a prairie dog for a meal. This went on till all of a sudden, there were hardly any prairie dogs. When this came to the attention of the Great Spirit of All Animals, he did something about it. He commanded the littlest owl of all, the tiny grey prairie owl, to live with the prairie dogs and to fly high over their towns, giving a shrill warning sound when he saw birds of prey. In appreciation the prairie dogs stored up seeds for the owls to eat in the winter. And that's why prairie dog towns usually have some little owls around.

In truth, burrowing owls feed on young prairie dogs, as well as insects, lizards, small

BLACK-TAILED PRAIRIE DOGS AND BURROWING OWL

snakes and mice. Owlets, in underground burrows, make sounds that closely resemble the rattlesnake's "rattle." Burrowing owl fledglings and their parents, gathered on a prairie dog mound after their burrow had been flooded, were observed by Lewis W. Walker:

> From afar I realized that they were waging a battle for possession of the burrow. Each time the rodent owner of the burrow came within about twenty feet of the mound, an adult owl hovered on rapidly beating wings. And at the slightest intimation that the prairie dog was going to make a shouldering rush to the entrance, the bird would swoop and rake its back. After several attacks from above, the prairie dog seemed to lose interest in repossession, and when I approached for a closer view, it sought refuge in a burrow many feet away.

A single shrill warning bark sounded in the town as a prairie rattlesnake slithered into a burrow. Prairie dogs plunged into their holes, for the rattler is a dreaded invader. Early observers thought that prairie dogs enjoyed the same harmonious relationship with rattlesnakes as they did with burrowing owls. Nothing could be further from the truth. Among the first to dispel this myth was the famous vertebrate paleontologist Edward Drinker Cope, whose searches for dinosaurs gave him many opportunities to investigate prairie dog towns. In his *Monograph on the Reptiles of North America* (1900) he writes:

> Prairie-dogs . . . seem to have a most intense dread of rattlesnakes. This little animal dreads not only its venomous bite, but more the loss of its young, which serve as food for these snakes that enter their burrows, take possession, and drive them from their homes. Where does one find a Prairie-dog town but that it is teeming with snakes and the strange little owl . . . that 'ducks' to passers in ludicrous solemnity? These do not constitute a happy family. The owls, though they generally occupy an abandoned hole or burrow, destroy the young Dogs.

E. C. Cates describes finding the dead, entwined bodies of a male prairie dog and a large bull snake. Apparently the snake had entered the prairie dog's burrow and the prairie dog had attacked it. The soil around the burrow indicated a long tussle between the two adversaries. The snake had been severely bitten in several places, and the prairie dog appeared to have been squeezed or choked to death. According to Cates, "A rattlesnake or bull snake that tries to get a square meal while in a prairie dog's burrow takes great chances of being severely mutilated or even killed."

Coyotes, reduced in numbers like the prairie dogs themselves, were once fearful enemies in the towns. The coyote's method is to hide behind a clump of vegetation or soil mound and depend on its protective coloration to make its still form unseen. When an unwary prairie dog appears, the coyote, with a quick rush, attempts to head it off and catch it.

More feared than the coyote is the badger, low-slung and powerful. Seeing a prairie dog flee into its burrow, or scenting a prairie dog in its den, the badger starts boring spirally into the burrow. He works at great speed, using long-clawed front feet, shorter-clawed hind feet and snout-like nose to bore down to his victim. In fifteen or twenty minutes the badger usually procures his prey.

Sometimes coyotes and badgers collaborate. J. Frank Dobie recounts stories of coyote-badger cooperation. Lacking the badger's digging prowess, coyotes have been seen flanking a digging badger. The badger got a prairie dog for his efforts, but not until two had come out past him and into the waiting coyotes' jaws. The badger reportedly makes use of the fleet coyote's ability to run prairie dogs to ground.

One of the handsomest and rarest of North American mammals is also a dreaded and relentless prairie dog foe. Creamy brown above, lighter below, with a dark mask across its eyes and four dark "boots," the Black-footed Ferret is long and slender enough to enter and traverse freely prairie dog burrows. Unless it can dig free through an escape tunnel, a prairie dog has little chance of evading a ferret in its

BLACK-TAILED PRAIRIE DOGS AND BLACK-FOOTED FERRET

burrow system. Always appearing in the vicinity of prairie dog towns, the Black-footed Ferret uses abandoned burrows for shelter and depends upon the communal rodents for his meals. Even when prairie dog towns were numerous on the plains these ferrets were seldom observed. Because of its habit of sitting up, with forefeet folded against its chest, its form at a distance is difficult to distinguish from the prairie dogs themselves. This dependence upon the prairie dog threatens the ferret's continued existence as part of the prairie fauna, and it is now listed as an endangered species.

Ravens and hawks also prey upon prairie dogs. Smaller animals, such as ticks and fleas, annoy prairie dogs. For relief they roll and squirm, chirring loudly, as they dust-bathe their thinly haired skins.

Around 1900 prairie dogs were more abundant than ever. Cattle grazing, which reduced the forage to favorable height, had increased the prairie dog food supply. The white man had killed off many of the prairie dog's natural enemies. Naturalists saw flourishing prairie dog towns, sometimes viewing them from train windows. C. H. Merriam, chief of the old-time Division of Biological Survey, found prairie dogs little disturbed by passing trains:

> I have often watched them from the 'Overland Limited,' some standing erect on their mounds; others chasing one another about or quietly feeding within forty or fifty feet of the roaring, rushing train, without showing the least outward sign that anything unusual was happening.

Of towns observed on foot Merriam also writes:

> When a person approaches a dog town the animals see him a long way off and keep a close watch on his movements. As he comes nearer an alarm note is sounded, at which those away from their burrows rush to the entrance mounds, where they sit or stand erect, nervously twitching their tails and chattering or barking excitedly. If he continues to move toward them the excitement increases, and most of the animals on the near side of the colony plunge headlong into their burrows. Some withdraw more slowly, and for some time their heads and eyes may be seen peering up from the funnel-shaped openings of the mounds. Those nearby are usually silent, while those at a little distance continue to scold and chatter.

According to Merriam's estimates prairie dog towns varied from a few acres in extent to colonies that covered thousands of square miles. One colony in Texas reportedly covered an area of some 25,000 square miles, measuring 250 miles by 100 or 150 miles, and boasted an estimated 400 million prairie dog inhabitants.

Such towns moved in waves across the plains. Merriam explains this gradual movement of prairie dog colonies:

> The ground immediately surrounding each burrow is usually cleared of small plants and kept clean and bare, and where burrows are near together the bare areas often join, so that in thickly populated colonies the ground is hard and

> smooth like a playground, and the animals are obliged to go some distance for food. This they dislike to do, lest they be pounced upon by enemies; hence, when the grass near their burrows has been consumed they dig new holes nearer the supply. It takes a long time for vegetation to regain a foothold on the hard floors of the dog towns, and the sites of old towns remain conspicuous for years after they are abandoned.

Theodore Roosevelt noted this phenomenon and realized the importance of the prairie dog in its natural community. Digging loosened deep layers and allowed the soil to be fertilized by droppings and nest material. Air and water containing solvents entered the soil where prairie dogs were active. Microbial life and small living things flourished on the oxygen and contributed to soil enrichment. Better growth of short grasses followed in the wake of prairie dog towns.

About 1880 man began to wage his relentless war against prairie dogs. Cattlemen regarded the prairie dog as a menace because of its consumption of grasses. Government experts estimated that thirty-two prairie dogs consumed as much grass as one sheep, 256 prairie dogs required as much grass as one cow. Of this war against the prairie dog Ernest Thompson Seton writes:

> During the first 20 years . . . (prairie dogs) scoffed at our best efforts. Their numbers increased, and their territory was widened. Arrows, guns, traps and Dogs were annoying trifles. But when the wise men of the Biological Survey came on the scene with their poison gas and deadly dopes the tide of battle turned; and one by one, the dog towns with their teeming, chippering potbellies, were left silent, and turned into irrigated fields and farms.

Although war was waged on the prairie dogs in the West, two pairs of Black-tailed Prairie Dogs were introduced on Nantucket about 1890. At first the increase in numbers was slow. But in 1899 Outram Bangs of Harvard counted 200 prairie dogs visible at one time. There were three or four towns, as well as smaller scattered groups living on the island. In 1900 the prairie dogs were deemed "a dangerous pest and nuisance, destroying crops and fields." Money was appropriated at a town meeting for the purchase of carbon bisulphide, and soon there was "not a dog left to tell the tale."

The female Black-tailed Prairie Dog and her young would soon feel chilling gusts of wind. Autumn rains soak the prairie. Silvery frost covers the grass in the mornings. At this time of year prairie dogs become very fat and are less frequently active above ground. Since they do not store food prairie dogs must "live on themselves" until spring. Soon the gray sky will lower and winter will come down from the mountains. With the first snow prairie dogs go underground, coming up briefly on sunny days. Blizzards heap drifting snow over the prairie dog town, and the animals stay below for long periods. Prairie dogs do not hibernate, nor do they store food in their burrows, yet they are able to survive the severest of winters. When a chinook follows a storm, warm air flows down from the mountains. The town becomes alive with prairie dogs feeding on tips of grasses exposed in the melting snow cover.

MEXICAN PRAIRIE DOG (*Cynomys mexicanus*)

Description. A more grizzled appearance, produced by numerous black hairs, grayish buffy coat color and more black on its tail, distinguish this prairie dog from its close relative the black-tailed.

Distribution. The Mexican Prairie Dog's range includes parts of three districts in central Mexico: Coahuila, San Luis Potosi and Nuevo Leon. In this species there are reported to be three molts per year. Other prairie dogs change their coats twice a year, a soft, heavily underfurred winter pelage replacing the thin summer hair.

Distribution of the Mexican Prairie Dog (right) and the White-tailed Prairie Dog (left)

WHITE-TAILED PRAIRIE DOG (*Cynomys leucurus*)

Description. The White-tailed Prairie Dog, and the two other species included in the subgenus *Leucocrossuromys,* are slightly smaller, more slender and shorter tailed than prairie dogs of the black-tailed group. Their tails are tipped or bordered with white.

Distribution. This species lives in the Rocky Mountain region of southern Montana, Wyoming, Colorado and Utah, at higher elevations than its town-dwelling black-tailed cousins.

Habits. Living in mountainous terrain makes it unnecessary for the White-tailed Prairie Dog to construct dikes to prevent flooding of its burrows. Over much of its range the White-tailed Prairie Dog hibernates to avoid the rigors of mountain winters.

UTAH PRAIRIE DOG (*Cynomys parvidens*)

Description. This species has a white-tipped tail and belongs to the White-tailed Prairie Dog group. Its fur is cinnamon or clay-colored mixed with buff or black-tipped hairs and is usually darker over the animal's rump.

Distribution. This species represents the westernmost outpost of prairie dog populations. It is also the least common and most restricted in its range of all the prairie dogs. Utah Prairie Dog towns are to be found over a fifteen-square-mile area on Packer Mountain in Utah, where their numbers are estimated at 1,500. Counts of nine towns totalled 2,775 prairie dogs of this species, now found in only five counties of south-central Utah. Its habits are similar to those of *Cynomys leucurus.*

Distribution of the Utah Prairie Dog (left) and the Gunnison's Prairie Dog (right)

GUNNISON'S PRAIRIE DOG (*Cynomys gunnisoni*)

Description. Cinnamon-buff coloration, overlaid with black hairs, marks the Gunnison's Prairie Dog. White tips and borders its tail, while underneath the fur is pale cinnamon to buff.

Distribution. From central Colorado into the southeast corner of Utah the Gunnison's Prairie Dog extends its range into Arizona and New Mexico.

Habits. More like ground squirrels in habits, Gunnison's Prairie Dogs have burrows that are scattered and a less communal way of life than the Black-tailed Prairie Dog's. Not only does this species neglect to build mounds, but it is also careless about allowing grass tufts, shrubs such as sagebrush and rabbitbrush, stumps and stones to remain near its burrow openings. Sloping grounds above the edges of meadows are favorite burrow sites. Using its front feet to loosen dirt, this species kicks soil from its burrow with its hind feet. The soil is not pushed with the front feet nor is any effort made to tamp the loose earth into mound shape. The relatively shallow burrows slant downward and lack the listening post or guard room always found in the plains prairie dog burrows. Pine needles, instead of dried grass, may be taken into the burrows for nesting material.

This mountain-dwelling prairie dog exists in numbers close to its original abundance, at least in parts of its range. Where slope exposure has encouraged tongues of foothills vegetation to creep up the mountainsides, the Gunnison's Prairie Dogs have extended their range, sometimes reaching the lush vegetation of mountain meadows or parks. Where they occur at higher elevations, young Gunnison's Prairie Dogs are known to have a higher growth rate during summer months. Possibly this is due to more abundant food supply. Certainly it is an adaptation to the shorter growing season of the mountains.

Although community spirit is less evident than it is among prairie dogs on the plains, Gunnison's Prairie Dogs nevertheless depend on each other for warning. When feeding they warily sit up on their haunches every five to ten seconds, looking out for danger. When alarmed they stand on toe-tips or climb on top of a rock or grass clump for a better look. A shrill warning bark causes general panic, as all prairie dogs sit upright attempting to locate the threat. Like all prairie dogs their lives depend on reaching the safety of their burrows in time.

Where prairie dogs can be seen today. Devil's Tower National Monument in Wyoming, Wind Cave National Park in South Dakota, Mackenzie State Park near Lubbock, Texas all have flourishing prairie dog towns. Prairie dogs may also be observed in Badlands National Monument (South Dakota), Hutton Lake National Wildlife Refuge (Wyoming), Theodore Roosevelt National Memorial Park (North Dakota) and Wichita Mountains Wildlife Refuge (Oklahoma).

GROUND SQUIRRELS

The genus *Spermophilus* includes some fifteen species that are well-adapted for life on the ground. They are short-legged; the four toes of their front feet and the five toes of their hind feet are provided with strong digging claws. Active by day, they search for their food of seeds, roots, bulbs, plant stems and leaves and insects. Numerous trips are made, with cheek pouches bulging, to their underground storage chambers. Summer is a time when ground squirrels become very fat. In the northern part of their range ground squirrels become dormant; in the southern part they may only avoid extreme weather conditions by staying in their burrows for short periods.

Whistling alarm calls and the habit of sitting upright for a better look around characterize these squirrels. They are less gregarious than the prairie dogs, although their colonies often attain large sizes, especially where the food supply is abundant.

Eight species of ground squirrels belong in the subgenus *Spermophilus*. Four of these squirrels—Uinta Ground Squirrel, Richardson's Ground Squirrel, Belding's Ground Squirrel and Townsend's Ground Squirrel—have unmottled fur. The other four—Arctic Ground Squirrel, Columbian Ground Squirrel, Washington's Ground Squirrel and Idaho Ground Squirrel—have spotted or mottled fur.

TOWNSEND'S GROUND SQUIRREL (*Spermophilus townsendii*)

Description. This relatively small ground squirrel is about nine inches in total length. Its tail is short, usually measuring less than one-third the length of the squirrel's head and body. The fur is soft, buffy or smoke gray washed with pinkish buff above and buffy white below. Reddish buff colors the sides of head and hind legs. The underside of the tail is cinnamon edged with white.

Distribution. Townsend's Ground Squirrels are found from south-central Washington through eastern Oregon, southern Idaho, parts of extreme eastern California,

Distribution of the Townsend's Ground Squirrel

Nevada and western Utah. They live in large colonies in dry, sandy, sagebrush valleys where their burrows are dug under sagebrush or out in the open. Sometimes Townsend's Ground Squirrels inhabit juniper-covered ridges where they burrow among lava rocks.

Occasionally other ground squirrel species overlap Townsend's Ground Squirrel's range. Where this happens different habitat preferences enable the ground squirrel species to live in the same area. Townsend's Ground Squirrels are abundant in dry sagebrush areas, while Richardson's and Belding's ground squirrels dig their burrows in adjacent meadowlands. Altitude usually separates the habitats of Townsend's Ground Squirrel and the larger Columbian Ground Squirrel.

Habits. Much of a Townsend's Ground Squirrel's life is devoted to being inactive. Climatic conditions such as high temperature and lack of moisture control its activity. From July to late January or February this squirrel is seldom seen above ground. In March the young squirrels are born, after approximately twenty-four days gestation. Then feeding and fattening become the concerns of the ground squirrels for the rest of the time they are above ground. June is harvest-time. Seeds of grasses, lupines and alfilaria ripen early in their parched habitat. From his observations of the Paiute Ground Squirrel (a subspecies of *Spermophilus townsendii*) J. R. Alcorn writes:

> In contrast to the males, which ordinarily require about 120 days to become fat, the females and juveniles are usually active about 135 days before they become fat enough to aestivate.
>
> The young of both sexes also fatten later than do the adult males, and, . . . usually go into aestivation about the same time as the females, namely, in the first week of July.
>
> Once these squirrels go into aestivation, they normally stay underground for seven and a half to eight months before emerging in the following spring. If, however, the food supply is not sufficient to fatten the squirrel in 120 to 135 days, it may stay out longer, or may emerge for a time in the autumn after only a short period of aestivation.

Environment and even microenvironment affect periodicity in ground squirrels. In Utah Townsend's Ground Squirrels usually are active until mid-July. But on an island in Great Salt Lake, where heat is absorbed by the lake during the daytime and radiated at night, accelerated growth of vegetation and ground squirrels was observed. By mid-June the ground squirrels on the island had gone underground.

The tunnels of a Townsend Ground Squirrel colony honeycomb the ground. Burrows are of three kinds: home or main burrows of adults, home burrows of young, and auxiliary burrows used by both adult and young ground squirrels. Home burrows of adult ground squirrels may tunnel underground for more than fifty feet, have two or more openings and go as deep as five feet below ground surface. Some of the burrows interconnect. Walter P. Taylor once excavated a Townsend's Ground Squirrel burrow, some thirty feet long, in Nevada:

> A nest was found in a large spherical cavity, so arranged that water could not have gotten into it. Fine straws made up the bulk of it, though white cotton twine had been very largely used to bind the straws loosely together. A couple of rags, a bit of rabbit fur, some wool, and a down feather were also incorporated into the nest.

A young squirrel's burrow also contains a nest made of grass or sagebrush bark, but has just one entrance and lacks the length and depth of an adult squirrel's burrow.

Auxiliary burrows, with no nest chambers, are seldom over five feet long. Usually excavated near feeding areas, they are used for refuge.

Like most ground squirrel babies, young Townsend's Ground Squirrels, eight to twelve in a litter, remain in the mother's burrow until they are about two-thirds grown. Alcorn describes their activity:

> The young may be seen thrusting their heads out of the burrow about May 1. After several days of viewing the world from the entrance of the burrow, they venture out, gradually increasing the distance they wander in search of food, until they are able to travel as much as a hundred feet and still find their way back home. Many of the young squirrels on their first few trips from the burrow get lost and start squeaking . . . . many of these lost young find their way back to their burrows unaided. . . . I never have seen any response from a female to the squeaking cries of the young.

Young Townsend's Ground Squirrels fall prey to prairie falcons, Swainson's hawks, red-tailed hawks and rough-legged hawks, and may be captured by gopher snakes and weasels. Badgers dig them out, even in winter when the squirrels are hibernating.

Because they feed chiefly on green vegetation, Townsend's Ground Squirrels often burrow into high ground near fields of alfalfa and growing grain. To get to better feeding areas these squirrels will travel a quarter of a mile or more, or readily swim across an irrigation canal. They also climb into bushes for food and nesting material and up onto leaning fence posts for a better view of their surroundings. They eat insects and ground squirrels of their own kind that are run over by cars.

The Paiute tribes, once dependent chiefly upon wild seeds and pinon nuts, have long used this ground squirrel for food. Alcorn gives the following account:

> In June of any year it is a common sight to see an old buggy, drawn by a small pony, laden with numerous containers for carrying water, one buck Indian and several squaws. Often the Indians have traveled for several hours from the Reservation to find a place where the squirrels are numerous and close to water. At this place the squaws carry water from a few to several hundred feet and pour several buckets-full into a burrow. The buck Indian catches, by the neck, the squirrel that is forced out of the hole. If the squirrel is thin it is usually turned loose; if fat it is either killed or placed alive in a sack or box. Sometimes the squirrels are eaten within a few hours after they are killed; but Indians often take live squirrels many miles to trade or sell to other Indians.
>
> In June, 1937, I came upon two Indians who had several hundred live squirrels in the back of an old automobile. These two men, about twenty-four years of age, and obviously of some education, said that they themselves did not eat the squirrels, but had caught them to take to Pyramid Lake, fifty-five miles away. There they would sell the squirrels to other Indians for ten cents each or three for twenty-five cents. These two Indians said that they had made from five to twenty dollars a day in this way. According to them the "old timers" were better customers than younger Indians.

> The younger generation of Indians who eat these squirrels usually dress them and use modern methods to cook them but the "old timers" do not. These old Indians bury the squirrels in hot coals. After the hair has burned off, the squirrel appears to be burned so much as to be unfit for food. However, when it is taken out of the fire one finds that only the skin is burned and that the meat and fat, the only parts eaten, are nicely roasted. I have eaten squirrels prepared in this way and in the modern way. Prepared by either method they are of good flavor and so fat as certainly to be highly nourishing.

Alcorn believes that the Indians may have had an effect on the distribution of this ground squirrel:

> It is possible that Indians for generations have taken squirrels of this species from one valley to another and there liberated them in an effort to insure a stable food supply. In this manner the Piute ground squirrel may have come to occupy a range of larger size than it normally would have had.

WASHINGTON'S GROUND SQUIRREL (*Spermophilus washingtoni*)

Description. This is a small, dappled ground squirrel. About nine inches in total length, its smoke gray fur is flecked with whitish spots. Reddish buff fur covers its hind legs, and its blackish-tipped tail is cinnamon colored, edged with white, on the underside.

Distribution. Low arid regions—grasslands, sagebrush, rocky slopes—in southeastern Washington and northern Oregon are the preferred habitats of this species. Howell quotes Vernon Bailey's 1896 field notes describing Washington's Ground Squirrel habits:

> They are most numerous along steep hillsides, in gulches, and in sagebrush along river bottoms. On the smooth, grassy prairie they are common and more evenly distributed. They collect where some protection is afforded by scattered bunches of sagebrush or *Chrysothamnus,* but avoid any dense cover from which they cannot look out on all sides.

IDAHO GROUND SQUIRREL (*Spermophilus brunneus*)

Description. This small, short-tailed ground squirrel has relatively large ears and a dappled grayish brown fur coat. Cinnamon or light brown wash is conspicuous on its back, which is sprinkled with small grayish white spots. Rufous fur covers its nose, outsides of hind legs and underside of tail. On its undersides the short, coarse fur is a gray-tawny color. Its tail has terminal hairs with five to eight alternating bands of black and white or tawny.

Distribution. Found in west-central Idaho, in the Weiser and Payette valleys, the Idaho Ground Squirrel's range meets or slightly overlaps that of the Townsend's Ground Squirrel.

Distribution of the
(A) Washington's Ground Squirrel
(B) Idaho Ground Squirrel

RICHARDSON'S GROUND SQUIRREL (*Spermophilus richardsonii*)

Description. This plains dweller, sometimes called the "Picket Pin" or "Flicker-tail," is a medium- to large-sized ground squirrel. Ten to fourteen inches in total length, it has a short tail and very small ears. Its fur is smoke gray washed or dappled with cinnamon-buff above and pale buff or whitish below. White or buff borders the tail, which is clay color or light brown on the underside.

Distribution. From the plains of Alberta, Saskatchewan and Manitoba this species ranges south into Montana, North Dakota, northeastern South Dakota, eastern Idaho,

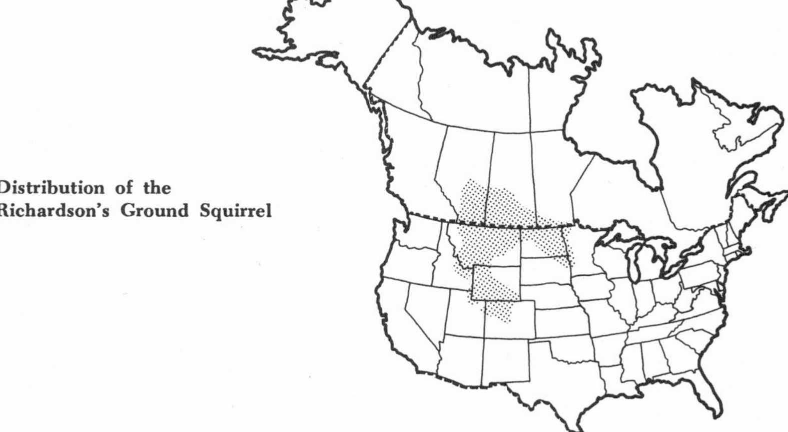

Distribution of the
Richardson's Ground Squirrel

Wyoming and nothern Colorado. A subspecies, *Spermophilus richardsonii nevadensis,* has an isolated range in northern Nevada, southeastern Oregon and southwestern Idaho.

Three other ground squirrel species have ranges that meet and overlap that of Richardson's Ground Squirrel. Hall and Kelson give the following characters that distinguish Richardson's Ground Squirrels from:

Belding's Ground Squirrel—by relatively and actually longer tail, more intense cinnamon pigmentation on nose and underparts and ochraceous buff rather than reddish on underside of tail.

Townsend's Ground Squirrel—by relatively and actually longer tail, cinnamon colored rather than whitish underparts, and tail with, rather than without, buffy white border.

Uinta Ground Squirrel—by the ochraceous buff instead of gray underside of the tail.

THE WAYS OF A RICHARDSON'S GROUND SQUIRREL

A "picket pin" stood erect on his hind legs as a car rushed by on the ribbon-like highway that crossed the Wyoming plains. Dropping to all fours, the picket pin gave a short, shrill whistle and darted toward his burrow. He paused on the gravelly soil of his burrow mound and then, with a tail flick, disappeared from sight.

Richardson's Ground Squirrels were common in this plains area. Large colonies occurred in some places, while stragglers lived here and there on the grasslands. Of these stragglers Eliot Coues writes:

> Every now and then, in odd out-of-the-way places, where there may not be another Gopher for miles perhaps, we come upon a solitary individual guarding a well-used burrow, all alone in his glory. The several such animals I have shot all proved to be males; and what is singular, these odd fellows are always larger than the average . . . peculiarly sleek and light-colored, and enormously fat. . . . I suppose they are surly old bachelors who have foresworn society for a life of indolent ease, though if I had found them oftener among their kind, I should have taken them for the Turks of the harem.

Ernest Thompson Seton once calculated a population of Richardson's Ground Squirrels in Manitoba to consist of "not less than ten per acre" or "5,000 to the square mile." Howell states that in places their burrows were once more numerous than prairie dog mounds, and estimates twenty ground squirrels to an acre. Coues describes the one-time abundance of these ground squirrels:

> It is one of the most abundant animals of our country, occurring by hundreds of thousands over as many square miles of territory, almost to the exclusion of other forms of mammalian life. Millions of acres of ground are honeycombed with its burrows. . . . I never saw any animals—not even Buffalo—in such profusion. I have ridden for days and weeks where they were continuously as numerous as Prairie-dogs are in their populous villages. Their numbers to the square mile are vastly greater than I ever ascertained those of *S. beecheyi,* the pest of California, to be, under the most favourable conditions.

Within its colony the picket pin occupied a relatively small home range. Rarely did he venture more than fifty yards or so from his burrow. He fed on grama grass heads, rabbitbrush blossoms and other flowering plants, and seeds of wild sunflowers, pig-weeds, bindweed and sagebrush. Occasionally he was able to catch a grasshopper, caterpillar or other insect. At the other end of the valley some of his relatives lived near a rancher's grainfield. Although many ground squirrels had been poisoned, some of the crop was destroyed every year. Strips along the edges of the field and bare areas surrounding burrows in the field revealed the ground squirrels' work. During spring planting time the squirrels dug up and ate many of the seeds. As the grain grew they consumed succulent stems, then later they pulled down the stalks and bit off the grain heads. Seton describes Richardson's Ground Squirrel depredations in a twenty-acre wheat field where there was:

> a perfect labyrinth of Ground-squirrel runs leading from all parts of the near prairie for 100 yards or more into the grain. The runs had no common plan beyond convergence at the crop; but each main run appeared to have on it a sort of refuge burrow at every 10 or 15 yards. These refuges differed from the residential burrows in being small, inconspicuous, half-hidden in the run, and without mounds. The Ground-squirrels would dodge from one to another, twinkling in and out of sight at the slightest alarm. If two happened into the same burrow, there was mischief brewed at once, and the weaker had to make a dash across country in search of some more hospitable retreat.

Other crops are sometimes raided. Near Laramie, Wyoming, Richardson's Ground Squirrels once invaded a large lettuce crop, nibbling on the heads and causing deformed lettuce growth.

Wherever they live Richardson's Ground Squirrels harvest seeds, carrying them off in cheek-pouch loads to eat and to store underground. A squirrel rescued from a hawk by Seton had 240 grains of wheat and almost 1,000 grains of wild buckwheat in its cheek pouches. As many as 269 grains of oats have been counted from a ground squirrel's cheek pouches.

During August harvesting time the Richardson's Ground Squirrel grew very fat. He spent less time sunning on top of the burrow mound and instead carried load after load of seeds down into his burrow. This hoard would furnish him food the following spring, after hibernation and before fresh spring grass sprouted. Mouthfuls of grass, for nest-building, were also carried down. His mate and five young were busy, too, carrying their stores underground.

Below ground a Richardson's Ground Squirrel burrow slants downward from the mound. Tunnels, about three inches in diameter, form a labyrinth three to six feet below ground and anywhere from twelve to forty-nine feet in length. In addition to the main entrance there is always a less obvious "back door" to the burrow. Most burrows contain a nest cavity six to nine inches in diameter. Burrowing owls and meadow mice sometimes are tenants in Richardson's Ground Squirrel burrows.

In the grass nest of her burrow the Richardson's Ground Squirrel's mate had given birth to eight babies in May. Late in June, when the young were a third their mother's size, they clambered up the burrow tunnel and peered out at their grassland world. During the summer days they were above ground much of the time. George Bird Grinnell writes of young Richardson's Ground Squirrels he watched at play:

> They were on the shelf of a flat block of weathered yellow sandstone in which there were several large cracks. They sat up on their hind legs, wrestled, pretended to bite necks and shoulders, and the one thrown down, when he fell, at once ran into a crack in the rock, to emerge a few seconds later and resume the game. We took it for granted then that this was a part of the sport, the defeated animal running away, and hiding for a moment to recover.

By the middle of August the Richardson's Ground Squirrel and most of the well-fattened squirrels that lived in the Wyoming plains colony came out above ground less often. Soon it was time for their long sleep underground—a dormant period that would last until early April. Further to the north these ground squirrels emerge a week or two later. Seton writes of Richardson's Ground Squirrels in Manitoba:

> This species appears above ground very regularly each year about the middle of April, without regard to the weather. Late snowstorms sometimes set in after its reappearance, and the Ground-squirrel becomes unenviably visible as it runs over the white, but this does not make it return to its winter sleep.

UINTA GROUND SQUIRREL (*Spermophilus armatus*)

Description. This eleven-inch-long, short-tailed ground squirrel has light brownish gray fur. Cinnamon fur colors its face and ears, and gray flecks the sides of its head and neck. Buff fur rings its eyes. Below the fur is pinkish buff or buffy white. Black hairs intermingle with buffy white hairs on its tail. The blackish tail distinguishes the Uinta Ground Squirrel from Townsend's Ground Squirrel (tail not blackish) and Richardson's Ground Squirrel (underside of tail clay color).

Distribution. Uinta Ground Squirrels are found in the foothills and mountains of

Distribution of the Uinta Ground Squirrel

Montana, Idaho, Utah and Wyoming.

Habits. Above ground only about five months each year, Uinta Ground Squirrels live in large colonies in mountain meadows nearly to timberline where their burrows are dug in soft soil, and in valley pasturelands and cultivated fields where they sometimes excavate along irrigation ditches. They prefer moist habitats with lush vegetation.

BELDING'S GROUND SQUIRREL (*Spermophilus beldingi*)

Description. Ten to twelve inches long, this ground squirrel usually has a broad brownish streak down its back. Its sides are grayish washed with buff. Pinkish cinnamon colors its forehead, tints the grayish underparts and is conspicuous on chest, forelegs, forefeet and hind feet. Its short, two- to three-inch-long tail is chestnut below, tipped with black and bordered with buff or white. Belding's Ground Squirrel is distinguished from Richardson's Ground Squirrel by its *shorter tail* that is *reddish* below rather than clay color, and from Townsend's Ground Squirrel by its *larger size* and *cinnamon-tinged,* rather than buffy or whitish *underparts.*

Distribution. Belding's Ground Squirrels are most often found in colonies in mountain meadows surrounded by coniferous forests. From eastern Oregon and southwestern Idaho their range extends south in the Sierra Nevada of California and includes northern and central Nevada.

Distribution of the Belding's Ground Squirrel

Habits. Because it often is seen sitting up very straight this ground squirrel is also known as the "picket pin." Belding's Ground Squirrels have high-pitched calls. Six or eight short, shrill notes are rapidly emitted to produce a trill-like sound. Grinnell and Dixon describe their calls:

> The usual call of warning consists of a series of from five to eight short shrill whistles uttered in quick succession and weakening toward the last. . . . A single note, *sirt,* is also frequently heard.

Colonies of Belding's Ground Squirrels are found within a surprisingly large altitudinal range. Of a subspecies, *Spermophilus beldingi beldingi,* in Nevada E. Raymond Hall states: ". . . it occurs from at least 9,000 feet elevation in the Sierra Nevada on the eastern side of Lake Tahoe, down to 4,850 feet elevation along the Carson River, . . ."

In their preferred mountain meadow habitat Belding's Ground Squirrels dig their burrows beneath stumps, logs or rocks or sometimes out in the meadow. Sometimes they remodel pocket gopher tunnels for their own use. Where they occur, Belding's Ground Squirrels are often found in large numbers. Linsdale once determined that some 500 Belding's Ground Squirrels lived in a twenty-five-yard-wide strip around the edge of a narrow, quarter-mile-long mountain meadow. Two acres in Klamath County, Oregon, were found to be home for some 466 Belding's Ground Squirrels. Sometimes these ground squirrels live in pastures, meadows or grainfields in valleys, where abundant food supports large numbers of them. An acre of Butte Valley, California, ranchland contained about 560 burrows. In many places "poison campaigns" have reduced or eliminated Belding's Ground Squirrel populations. Belding's Ground Squirrels avoid steep, rocky slopes, dense timber or brush and marshy ground.

Young Belding's Ground Squirrels, usually born in April, come above ground in May when they are about one-third grown. Litter sizes vary from four to twelve, but commonly number seven or eight babies. As they forage the young are constantly warned by low-pitched, double barks from the mother squirrels.

Stems and seeds of grasses and leaves and flowers of plants are food for these squirrels, supplemented by an occasional insect. Green food, abundant during the summer in mountain meadows, enables Belding's Ground Squirrels to remain active above ground longer in the year than ground squirrels living in dry valleys at lower elevations. Arthur H. Howell writes:

> The date of entering hibernating quarters varies with the dryness of the summer and the supply of green vegetation. During some seasons, in the valleys, most of the squirrels disappear by July 10, but in seasons of greater rainfall they remain out several weeks longer; in the mountains, small numbers may be seen above ground as late as the first of September. In spring they begin to appear about the middle of February and by the first week in March are usually out in force, even if obliged to burrow through a foot or more of snow to reach the surface.

In the Sierra Nevada Belding's Ground Squirrels go into hibernation by late September and emerge the following April.

COLUMBIAN GROUND SQUIRREL (*Spermophilus columbianus*)

Description. Grayish buff or brownish fur above, flecked with small buffy spots, and reddish fur on face, legs and tail distinguish this large ground squirrel, thirteen to seventeen inches in total length. Below the fur is buff or tawny and the sides of the neck are gray.

Distribution. Grasslands and open timber are home to the Columbian Ground Squirrel. Its range includes the eastern part of British Columbia, extreme western

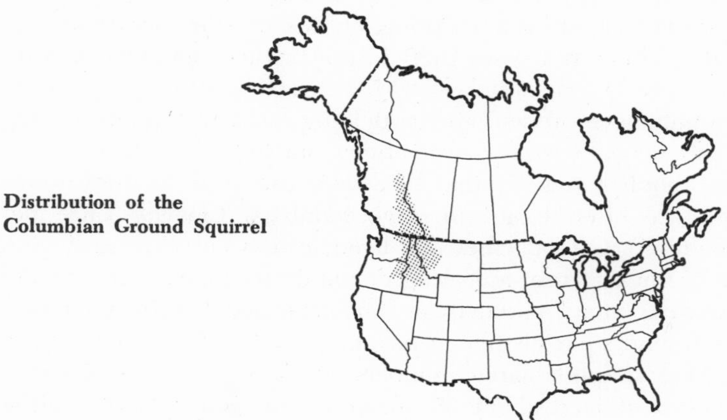

Distribution of the Columbian Ground Squirrel

Alberta, eastern Washington, northeastern Oregon, Idaho and western Montana.

Habits. Studies of the Columbian Ground Squirrel by William T. Shaw make its life history better known than those of other ground squirrels in this group.

During a five-year period Shaw excavated 136 dens. Most of the dens were located on gently sloping, south-facing hillsides, but some were on north-facing slopes. Variations in den construction correlated with two factors: 1) age and size of the ground squirrel occupant and 2) ground surface in which the den was located. An adult Columbian Ground Squirrel's hibernation den, located in a flat field, is elaborately built and furnished with a drain. The eight to ten inch diameter nest chamber or cell is about forty inches below ground and contains a grass nest. A younger squirrel's hibernation den is casually constructed, perhaps because its builder has been too busy above ground during the first months of life. The six-inch-diameter cell may be no more than six or eight inches below ground. Flatland hibernation dens of older squirrels usually have one or more drains, simple tunnels slanting downward in front of the nest chamber. These divert any water that floods or seeps into the burrow and thus the nest is kept dry. Hibernation dens that tunnel into hillsides lack these drains. Instead the nest chamber is protected by the slight ascent the tunnel makes as it leads into the chamber.

Columbian Ground Squirrels spend seven to eight months each year in their hibernation dens. Usually the hibernation dens are tunnelled out from the summer dens, but sometimes a squirrel makes a separate excavation. Each squirrel seals itself into its hibernating cell by tamping a two-foot-long earth plug into the burrow. In the lowland valleys and on the prairies, where vegetation has ripened and no longer supplies the needed moisture, the fattened ground squirrels become increasingly lethargic. As early as late July or early August they go below ground.

Sealed in its hibernating cell a Columbian Ground squirrel sits back on its hips, curls its head and body forward and presses its nose against its belly to form a loop. Within a few hours breathing and heartbeat have slowed almost to a stop. The squirrel's body temperature drops from 98 degrees Fahrenheit to 40 degrees, and it passes the long seven or eight months oblivious to the world outside the hibernating cell.

In the mountains some squirrels are active during September and young squirrels may be above ground as late as the first week in October. In the gentle valleys near Pullman, Washington, Shaw found that squirrels with dens on southwest slopes retired, emerged from hibernation and bred six to ten days earlier than squirrels living on northeast slopes.

Small food caches are found in the hibernating dens of many of the older male Columbian Ground Squirrels. Wild seeds or bulbs, tucked into the mulch at the rear of the hibernating cell, are used for food when the males awaken in spring, a week or ten days before the rest of the squirrel population. If the snow cover is melting and the weather not stormy, the safeguard cache may be left untouched in favor of sprouting green plant food.

By late February or March Columbian Ground Squirrels stir themselves. Soon after digging to the surface and relocating their summer dens, mating occurs. The squirrel colony is a place of frenzied activity at this time. After twenty-four days gestation litters of two to seven pink, hairless ground squirrels are born in many of the colony's dens. Until they are four weeks old the young remain in the nest. Their growth is so rapid that by the thirtieth day the youngsters are ready to fend for themselves and soon dig their own dens. Rapidly repeated sharp chirps, the squirrels' alarm calls, are almost constantly heard when the young venture above ground in May.

Of the habitat preference of the Columbian Ground Squirrel Howell writes:

Apparently the animals prefer rough, rocky, half-forested hillsides, but in many places they are numerous in hay meadows, grainfields, stony pastures, and open pine flats.

Narrow paths lead from burrow to burrow and radiate from the dens to feeding grounds used by the colony.

Summer dens are complicated structures usually excavated beneath logs, stumps or boulders. Vernon Bailey writes of a summer den he dug out in Glacier National Park, Montana:

> The mound at the entrance of the burrow contained about four bushels of earth and stones brought from the burrow, and the lower part was packed and hard as though an accumulation of several years. There were two other openings farther back from which no earth had been thrown and evidently they had been tunneled to the surface from below. The main shaft of the burrow was usually 3 or 4 inches in diameter, and back a couple of feet from the entrance, just before the burrow forked into two main shafts, was a roomy chamber where the squirrels could turn around and sit up comfortably, a sort of reception room. Near secondary forks were also two other chambers which may have served several purposes, such as convenience in storing earth brought out of the tunnels, or places of retreat from which to watch for enemies that might enter the burrow from either direction. Well back about 8 feet from the entrance and a foot below the surface of the ground was a large nest chamber nearly filled with old soft nest material. . . . At the bottom it was damp and moldy, but from the bed in the center to the top it was dry and clean, and a few fresh, green blades had been brought in for food or nest material. It had evidently served as winter quarters for the old squirrel and as a nest for her young and was being prepared for the coming winter. From one side of the nest chamber the burrow led down to an older and deeper chamber of some previous year, containing at the bottom an old rotten nest half full of excrement. A tunnel ran from it back toward the main entrance and into the main tunnel near the middle, making an easy way of escape if an enemy should dig to the first nest. Back of the nest a small shaft led to the surface of the ground and another opened out at the end of the first main fork of the tunnel. These rear openings were half concealed in the grass and evidently were for use as avenues of escape in case the burrow should be entered by a weasel or dug out by bear or badger.

Food for Columbian Ground Squirrels consists of leaves and stems of many kinds of grasses and herbs. Bulbs of glacier lilies, camas and wild onions are eaten. Buttercups, lupines and dandelion flowers, and currants, gooseberries and serviceberries are relished. Sometimes the ground squirrels become agricultural pests. On a 200-acre stretch of land some 1,200 squirrels were trapped, an average of six squirrels per acre. It was estimated that these squirrels consumed enough grass to support twelve sheep or three cows.

ARCTIC GROUND SQUIRREL (*Spermophilus undulatus*)

Description. Reddish brown fur, abundantly flecked with whitish spots covers the Arctic Ground Squirrel's back. Its head is tawny or cinnamon colored. Tawny fur of the underparts gives way to buff or grayish white in winter, and the winter pelage in general appears much more pale. This is a very large ground squirrel, eleven and one-half to nineteen inches in total length.

Distribution. The Arctic Ground Squirrel is the only squirrel species that occurs in Asia as well as in North America. Its North American home includes much of Alaska, and islands off its coast, and the Yukon territory and districts of MacKenzie and Keewatin in Canada.

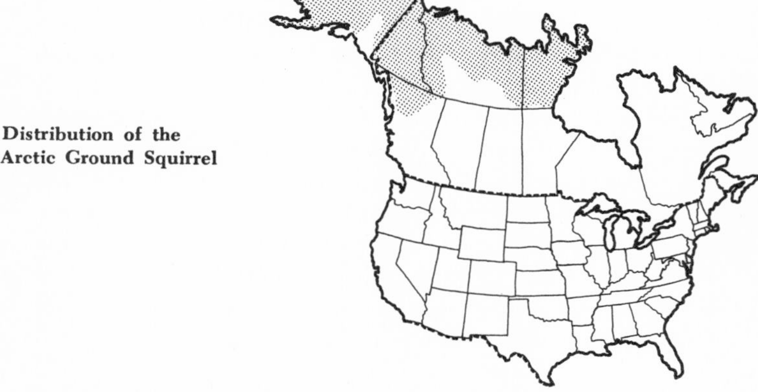

Distribution of the Arctic Ground Squirrel

Habits. In 1821 Captain Parry, in His Majesty's ships "Fury" and "Hecla," sailed from England to make a second search for a northwest passage from the Atlantic to the Pacific. John Richardson gives one of the earliest accounts of Arctic Ground Squirrels observed on the Arctic coast:

> The grey Arctic marmot is common in stony barren tracks, but delights chiefly in sandy hillocks, amongst rocks, where it burrows, living in society.

Edward A. Preble observed Arctic Ground Squirrels living in burrows in clay banks, above the high-water mark, along the Mackenzie River. Gravelly ridges were their home sites on the Barren Grounds near Cape Eskimo, Hudson Bay. In northern British Columbia Preble saw these squirrels in broad valleys covered by shrubby vegetation and on gentle lower slopes of mountains. Where they are not hunted by Eskimos, Arctic Ground Squirrel colonies are large and compact, and individuals are less wary.

Howell writes that Olaus J. Murie excavated an Arctic Ground Squirrel burrow in McKinley National Park, Alaska. The burrow had two openings, many branches, and went twenty-seven inches below ground. In the park the ground squirrels were numerous in the valleys, on mountain slopes and ridges. Their burrows were excavated among willows, in the plant cover of the slopes or among rocks.

ARCTIC GROUND SQUIRREL
(*Spermophilus undulatus*)

Studying Arctic Ground Squirrel habitat near Point Barrow, Alaska, William V. Mayer writes:

> At first glance the flat arctic tundra appears a distinctly unfavorable habitat for a burrowing animal. The ground of the arctic slope is permanently frozen to a depth of many feet throughout the year, except during the brief summer, when it may thaw superficially to a depth of a few inches or feet. . . . The high permafrost table greatly restricts the available habitats. Even in the arctic, however, a relatively low permafrost table exists in certain areas correlated with good drainage, and within these restricted regions ground squirrels have a suitable habitat. . . . the squirrels may locate their burrows in the ridges and river sandbanks, hillocks, and other raised areas with a sandy soil, sandy loamy soil, or sandy clay soil.

Numerous small bodies of water dot the tundra in summer. Arctic Ground Squirrels readily swim to reach foraging areas. Mayer observed squirrels travelling as far as 1,500 yards from their home burrows. Because of the lack of cover on the tundra Arctic Ground Squirrels crawl over the ground. Mayer believes this "tundra glide" renders them less conspicuous. When frightened, the squirrels bound away to their home burrows.

Mayer found burrows of Arctic Ground Squirrels to go no deeper than three feet, eight inches inches below ground. A colonial burrow that was excavated had sixty-eight feet of tunnel on three levels and six openings. Side passages serve as toilets so tunnels and nest are kept clean. A nesting burrow that was excavated had three openings and two nests. Sometimes a nesting chamber is located on a side passage of the burrow and has two entrances, or it may be situated at the end of the main tunnel. Dry grass, plus small amounts of lichens, ground squirrel fur, caribou fur and green leaves are used for nesting material. Peripherally located "bachelor burrows" were found to have single openings, tunnels up to ten feet long and no nests.

Plant food is the staple of the Arctic Ground Squirrel's diet. Richardson writes: ". . . their pouches were observed about the middle of June to be filled with the berries of *Arbutus alpina* and *Vaccinium vitis-idaea,* which were just then laid bare by the melting of the snowy covering under which they had lain all winter." *Polygonum* and *Astragalus* seeds are eaten, as well as mushrooms, caterpillars, beetles, insect larvae and pupae. Caribou antlers are gnawed upon by Arctic Ground Squirrels.

When foraging, an Arctic Ground Squirrel explores the feeding area, nervously running from place to place, stopping and sitting up, then seizing a plant with one or both forepaws and biting off the top. Often when its cheek pouches are filled a squirrel will stuff its mouth with dry grass to carry back for its nest. The ground squirrels often are seen resting on rocks near Hoary Marmots, depending on the marmots' keen sight and hearing to warn them of danger. Enemies of the Arctic Ground Squirrel are ermine, grizzly, wolf, wolverine, and gyrfalcon.

Arctic Ground Squirrels are sometimes carnivorous, feeding on dead of their own kind, as well as carcasses of such birds as murres, puffins, gulls and snow buntings. Around mining camps the squirrels are notorious beggars, relishing foods with high fat content. Eskimos use Arctic Ground Squirrels for both food and clothing. Because of the double-noted call, "keek-keek," the Eskimo name for the Arctic Ground Squirrel is "sik-sik."

During summer there is constant light over much of the Arctic Ground Squirrel's range, yet the animals appear to respond to a regular diurnal light rhythm. Mayer observed that the squirrels go underground for at least seven hours at night. Shortly after 9 P.M., influenced probably by the lower intensity of light, the squirrels went to bed. Around four the next morning the squirrels became active again.

Young Arctic Ground Squirrels are born in June, grow rapidly during the brief summer and by October are ready for hibernation. Noting that the conditions under which a mammal such as the Arctic Ground Squirrel lives place certain demands upon the growth rate of its young, Richard G. Van Gelder writes:

> An Arctic hibernator, such as the Arctic ground squirrel, must attain a sufficient amount of weight, including fat, within fewer than five months after birth in order to survive a seven-month hibernation period. . . . By the time the young ground squirrels are a month and a half old they are four-fifths the size of adults . . . at four months . . . they have attained adult size and weight and have enough fat to last them through hibernation.

Four ground squirrel species are grouped in the subgenus *Ictidomys.* Three species—Thirteen-lined Ground Squirrel, Mexican Ground Squirrel and Spotted Ground Squirrel—have upperparts with nearly square white spots, while the fourth species, the Perote Ground Squirrel has clay color upperparts.

THIRTEEN-LINED GROUND SQUIRREL (*Spermophilus tridecemlineatus*)

Description. Thirteen whitish stripes, some continuous and others interrupted to form rows of spots, identify this ground squirrel. Sometimes it is called the "federation squirrel" because of its stars and stripes; actually there may be a few more or less than thirteen stripes and many more stars or spots. Stripes of blackish or dark brown and whitish alternate along the back from head to rump. Each dark dorsal stripe has a row of almost square white spots. The stripes on the sides of the body are poorly defined. Cinnamon or buffy color gives way to pale buffy, pinkish or yellowish white underparts.

This small- to medium-sized ground squirrel, seven to eleven and three-quarter inches in total length, also goes by the name Striped Ground Squirrel. Its head is long and narrow, with large eyes and small, low-set ears; its body shape is slender, and its tail is short-haired and about one-half the length of the squirrel's head and body.

Audubon writes admiringly of the Thirteen-lined Ground Squirrel:

> In the warm days of spring, the traveller on our Western prairies is often diverted from the contemplation of larger animals to watch the movements of this lively little species. He withdraws his attention from the bellowing Buffalo herd to fix his eyes on a lively little creature of exquisite beauty seated on a diminutive mound at the mouth of its burrow. It darts into its hole, but, concealed from view, and out of the reach of danger, its tongue is not idle. It continues to vent its threats of resentment against its unwelcome visitor by a shrill and harsh repetition of the word *seek-seek.*

Distribution. The Thirteen-lined Ground Squirrel is typical of the Great Plains and prairie regions of central United States and Canada. From Alberta, Saskatchewan and Manitoba its range extends east as far as central Ohio, west into northwestern Utah and eastern Arizona and south through east-central Texas to the Gulf coast.

Land that has been cleared for farming has opened new territory for Thirteen-lined Ground Squirrels, and their range has expanded, especially eastward, during the last century. Where they are abundant Thirteen-lined Ground Squirrels may number from five to twenty per acre. Soil type, determining the type of vegetation and the ease of burrowing, affects the distribution and abundance of these squirrels.

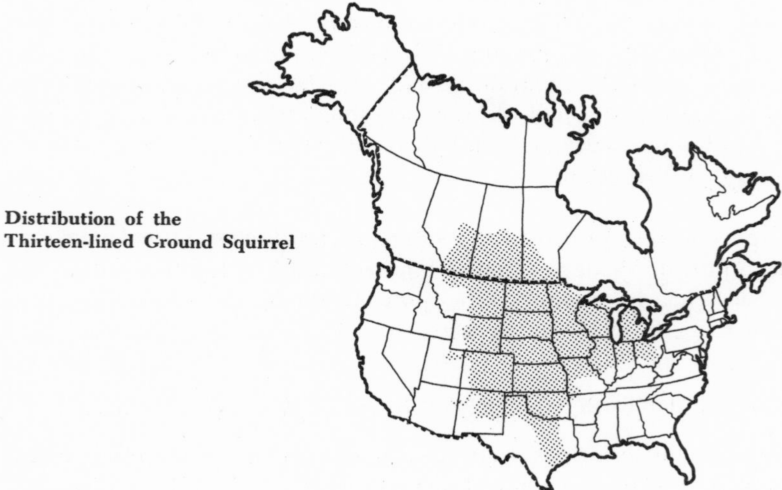

Distribution of the Thirteen-lined Ground Squirrel

THE WAYS OF A THIRTEEN-LINED GROUND SQUIRREL

The Thirteen-lined Ground Squirrel lived among others of his kind in a flat, open grassland area in Missouri. The entrance to his burrow, well concealed by grass, opened into a vertical tunnel which dropped several feet, then made a right-angle turn and zig-zagged horizontally for nearly twenty feet with numerous side pockets, storage areas for food, several passageways leading to the ground surface and an oval nest chamber, about nine inches in diameter, containing a nest of fine grass and rootlets. Leading downward from the nest chamber was a drain, protection against water in the burrow. The extent of the burrow indicated its occupant was a mature ground squirrel. Younger animals dig shallower, less elaborate dens that are six or more feet in length and close to the ground surface.

Earth plugs sealed the side entrances of the ground squirrel's burrow. At night he was careful to tamp earth or grass into the main entrance, too. Unlike many ground squirrel homes, a Thirteen-lined Ground Squirrel's burrow is not marked by a soil mound. When he excavates his den the squirrel fills his cheek pouches with earth which he carries away some distance to scatter. This made his doorway hard to find.

Emergency shelters, tunneled beneath the sod, were often used by the ground squirrel. A maze of small pathways connected these tunnels, near favorite feeding areas, with his main burrow. A soaring hawk or stalking fox or coyote sent the foraging squirrel flashing into one of his tunnels.

In spite of the fact that they dig their burrows in close proximity, Thirteen-lined Ground Squirrels are generally antagonistic toward each other. During most of the year the male ground squirrel led a solitary life and stayed within his acre-and-a-half home range.

Bright sunlight brought the Thirteen-lined Ground Squirrel out of his burrow about nine o'clock every morning. When days were overcast he usually remained in his burrow. Except for brief underground naps, he could be seen scampering about until late afternoon. Most of this time he spent foraging and carrying off foods for his underground storehouses, much of which would never be eaten. The bird-like, high-pitched, trilling call of the Striped Ground Squirrels sounded frequently over the grassland; their alarm call, heard less often, was a short, sharp whistle.

Always the squirrel kept a sharp watch for enemies. Occasionally he stood bolt upright on his haunches, forelegs pressed against his body, hind legs and tail forming a tripod to support his straight back, and surveyed his terrain for several minutes at a time. Whenever the ground squirrel was startled he ran into his burrow, only to turn around and pop his head out for another look.

As the squirrel fed, the alternating pale and dark stripes interrupted his body outline, making him almost indiscernible in the midst of grass stems, seed stalks, earth patches and flickering shadows.

Coming upon some ripened grass seeds the Thirteen-lined Ground Squirrel first filled his cheek pouches. He stuffed the seeds into his mouth with his forefeet, then lowered his head so that contraction of neck muscles forced the seeds into his cheek pouches. Then, a few at a time, he let seeds slip from his cheek pouches into his mouth for chewing. He discarded the husks and swallowed the seeds. During the summer he took many cheek-pouch loads into his burrow for storage. As many as 23,000 unshelled oat kernels were estimated in one ground squirrel's nest cavity. Other burrows have been found to contain more than 2,000 wheat kernels. Cahalane lists

THIRTEEN-LINED GROUND SQUIRRELS
(*Spermophilus tridecemlineatus*)

goosefoot, wild sunflower, knotweed, gromwell, ragweed, black locust, cotton, dandelion, buffalo-bur, domestic flax, wild beans, cultivated grains and many kinds of grasses and legumes as food for Striped Ground Squirrels.

At least half the Thirteen-lined Ground Squirrel's diet consisted of animal material. Vernon Bailey examined stomach contents of eighty Striped Ground Squirrels and found that half the contents were insects—grasshoppers, crickets, caterpillars, beetles, ants and insects' eggs. Butterflies too, are sometimes captured and eaten, as well as an occasional bird egg, fledgling or young mouse. Grasshoppers were favorite prey of the Thirteen-lined Ground Squirrel. Pursuing a grasshopper with swift, cat-like leaps, he pinned it down with forefeet, then bit into its head. Caterpillars were tackled in a different way. The squirrel pounced on their bristly bodies, striking repeated blows with the claws of his front feet.

Insect eating by Striped Ground Squirrel colonies compensates for their raids on newly planted seeds and sprouting crops on nearby cultivated lands. Most of the damage the squirrels inflict on farmers' fields occurs in springtime, when few insects are available. Seldom do the squirrels venture far out into cultivated fields. Their predilection for insects prevents grasshoppers and other insects from multiplying into hordes and makes them beneficial inhabitants of meadows and grasslands.

By late summer the Thirteen-lined Ground Squirrel became noticeably fat. His faded, worn pelage had been replaced by bright new fur. Like other fat squirrels he became irritable and drowsy. In mid-October he was one of the first squirrels in the colony to retire below ground. Many of the younger and thinner ground squirrels would still be active, accumulating fat on their bodies, for three or four weeks more. The squirrel plugged the openings of his burrow and began his long sleep. His body curled into a ball, nose touching belly, tail over head. As he became torpid his body processes slowed—heartbeat dropped from a normal of about 200 per minute to an average of seventeen beats. Respiration was reduced to as low as seven per minute; oxygen consumption decreased to about seven percent the amount he used when active. Gradually his body temperature sank from 86–106 degrees Fahrenheit to as low as 37 degrees. Every few weeks during the long months underground the squirrel awoke and stirred, then resumed hibernation. Severe cold, with freezing temperatures penetrating deep into the soil, aroused the ground squirrels from dormancy several times during the winter. This caused their body temperatures to rise and prevented their freezing to death. Unusually warm temperatures also awakened the squirrels now and then during their hibernation.

The first signs of the squirrel's spring awakening occurred one day in late March. His respiration rate increased and he humped his body. Slowly his body unrolled, but for a time he could not raise his head. The front part of his body appeared to wake more rapidly than the rear. Then his eyes opened and his forefeet supported his body. But sleep overcame him; in the gradual process of awakening hibernation usually passes into sleep. When he awoke again the squirrel found the ground thawed enough so that he could dig his way out. His stored body fat was used up and he weighed just over half his pre-hibernation weight of nine ounces. Eagerly he ate from his stores of seeds, nibbled at new green plant sprouts and cleaned out his burrow.

Two weeks later the females began to emerge from hibernation. Four weeks after mating, about mid-May, there were litters of ground squirrels in many of the colony's burrows. Large litters of eight, nine or ten babies were common, although a few nests contained as few as four and one had fourteen. Important as food for predators, Striped Ground Squirrels compensate for their losses by producing large families.

One female Thirteen-lined Ground Squirrel gave birth to seven babies in a Nebraska museum. She had arrived the day before, packed into a crate containing elephant bones and hay. Marjorie Shanafelt describes "Streak's" babies:

> When the babies were one day old they could prop themselves up on their forelegs, but the hind legs were absurdly weak. They lifted their too-large heads sturdily, and could, by wriggling and squirming in a lively manner, cover considerable ground . . . By the end of the week their backs were growing quite dark and one of them showed the first indications of tiny whiskers on his diminutive nose.

Downy fur covers the head, shoulders, back and sides of the baby squirrels, and by the twelfth day their stripes are visible and the babies trill noisily in the nest. About two weeks later their eyes open. At four weeks "Streak's" seven offspring were "gloriously dressed in polka dots and 13 rich brown stripes with tawny lines in between . . . Along their sides long hairs stuck out giving them that delightfully immature look so charming in kittens and young birds."

Soon the babies were ready to be weaned. When they appeared two-thirds grown, and shortly after their first ventures from the home burrow, the young ground squirrels dispersed and dug their own small, shallow dens nearby.

MEXICAN GROUND SQUIRREL (*Spermophilus mexicanus*)

Description. Nine longitudinal rows of nearly square light buff spots over its brown back distinguish this medium-sized (eleven and one-quarter to twelve and one-half inches in total length) ground squirrel. Its long, slightly bushy tail has buff-tipped hairs. Feet, sides and underparts are white to pinkish buff.

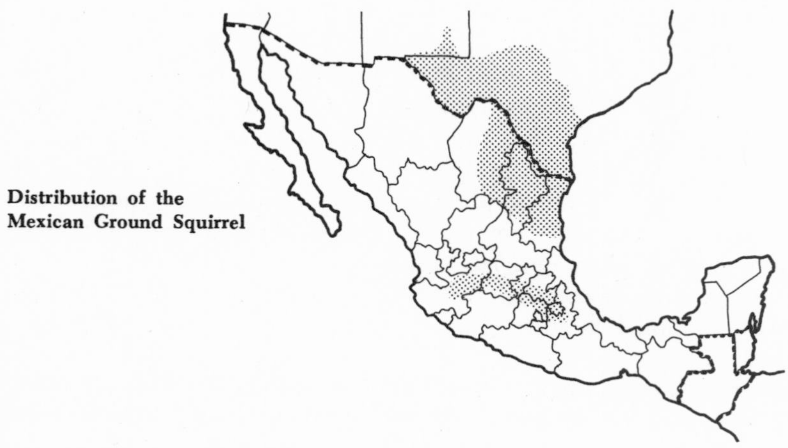

Distribution of the Mexican Ground Squirrel

Distribution. Desert areas in southeastern New Mexico and southwestern Texas, as well as in the Mexican states of Coahuila, Nuevo Leon and Tamaulipas, are home to one subspecies, *Spermophilus mexicanus parvidens*, sometimes called the Rio Grande Ground Squirrel. Another subspecies, *Spermophilus mexicanus mexicanus*, occurs in the central Mexican states of Jalisco, Guanajuato, Querétaro, Hidalgo, Distrito Federal and Puebla.

Habits. Mexican Ground Squirrels dig their burrows on sandy plains, usually locating them at the base of a mesquite or other bush or near the fringe of a cactus.

Seeds and insects are eaten in approximately equal amounts. Mesquite beans and acacia seeds are favorite foods. Like their thirteen-lined relatives these squirrels sometimes feed in cultivated fields of corn, grains and other crops.

Except for a shrill, whistled call when alarmed, Mexican Ground Squirrels are quiet and shy. Their burrows are plugged when they are "at home," and though these squirrels do not hibernate they stay in their burrows for considerable periods when the weather is cold.

SPOTTED GROUND SQUIRREL (*Spermophilus spilosoma*)

Description. The Spotted Ground Squirrel is small (total length seven to nine and a half inches), and its grayish brown or reddish brown coat is marked with squarish and rather indistinct spots of white or buff. Its short, round tail appears slightly bushy toward the tip. Its underparts are whitish.

Distribution. Dry, sandy soils from the southern part of South Dakota, Nebraska, eastern Colorado and western Kansas, Arizona, New Mexico, the Panhandle region, western Texas, south into the Mexican states of Chihuahua, Coahuila, Nuevo Leon, Tamaulipas, Durango, Zacatecas and San Luis Potosí are home to the Spotted Ground Squirrel.

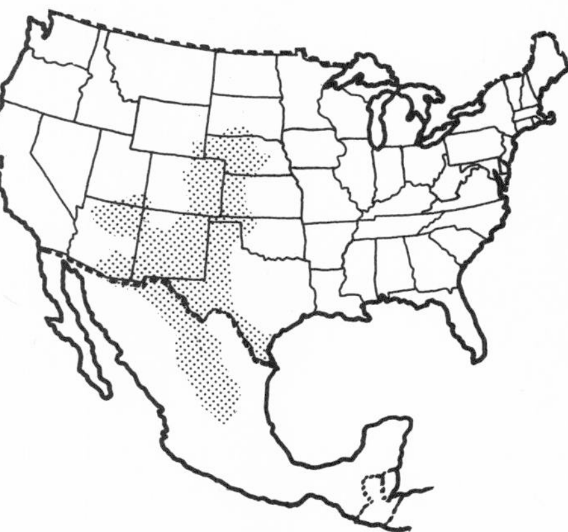

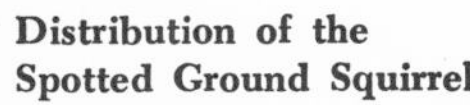
Distribution of the Spotted Ground Squirrel

Habits. Spotted Ground Squirrels live in drifted sand along river flats, in grassy parks, in open pine forests and less frequently on rocky mesas. Usually a bush, weed patch or rock protects the entrance of a Spotted Ground Squirrel burrow. Sometimes their burrows are renovated kangaroo rat homes. Howell describes an excavated Spotted Ground Squirrel burrow that had three openings, about twelve feet of tunnel and went no deeper than eighteen inches below the ground. At the end of the tunnel was a small, round chamber containing a grass nest.

When above ground these shy squirrels never venture far from their burrows. Their call is a "long, bubbling, birdlike whistle" or trill. They run and stop, bodies pressed to the ground, tails held low, much as a lizard moves. Their diet includes cactus pulp, mesquite beans, saltbush seeds, sandbur, sunflower, gourd, grasshoppers and beetles.

Spotted Ground Squirrels probably do not hibernate in the southern parts of their range; their northern relatives are believed to hibernate during severe weather.

PEROTE GROUND SQUIRREL (*Spermophilus perotensis*)

Description. The Perote Ground Squirrel resembles the Spotted Ground Squirrel but is larger (nine or ten inches in total length), and has a shorter tail. Small white indistinct spots cover the yellowish-brown fur on its back. Blackish fur marks its head; underparts are buff-colored.

Distribution. Restricted to the high plains at the extreme eastern border of the Mexican tableland of Puebla and Veracruz, Perote Ground Squirrels frequently live around the edges of cornfields and wheatfields.

FRANKLIN'S GROUND SQUIRREL (*Spermophilus franklinii*)

Description. This large, thirteen and three-fourth to sixteen and a half inches in total length, ground squirrel is placed by itself in the subgenus *Poliocitellus*. Relatively slender limbs and body give the Franklin's Ground Squirrel a tree squirrel-like appearance. The fur on its head is slightly darker than that of the upperparts; the sides are lighter. Brownish gray, speckled with black, the fur has a barred appearance, especially noticeable on the rump where the fur has a yellowish wash. Its relatively long tail (more than half the length of head and body) is blackish mixed with buff and bordered with creamy white. Dark gray fur covers its feet; the underparts are yellowish white, gray or buff.

Named in honor of Sir John Franklin, leader of the Overland Expedition of 1818–1822, on which the species was discovered, Franklin's Ground Squirrel is also known as the "gray gopher" and the "whistling ground squirrel." Of the Franklin's Ground Squirrel's call Seton writes:

> Most of the Ground-squirrels are noted for the great variety of the sounds they produce, but this is the musician of the family. It utters the same calls as the others, but expresses them in a fine, clear whistle. Its ordinary note . . . is in a high degree musical, resembling the voices of some of our fine bird-singers . . .

Distribution. From central Alberta, Saskatchewan and southern Manitoba Franklin's Ground Squirrels are found in the central United States as far south as Kansas, Missouri, Illinois and western Indiana.

Within its geographic range the Franklin's Ground Squirrel varies greatly in numbers. In the prairie regions of Missouri Charles W. and Elizabeth R. Schwartz write that the species "is generally rare but may be common locally." Hall states that in eastern Kansas "where one is seen a search will reveal six to ten others within a radius of a quarter of a mile."

Franklin's Ground Squirrel populations fluctuate—periods of their abundance alternate with years of scarcity. Lyle K. Sowls estimates that in Manitoba peaks in Franklin's Ground Squirrel populations occur every four to six years. When populations are dense these squirrels number five to thirty per acre. Then natural controls, disease and parasites, bring about sharp declines in numbers of the squirrels.

In 1867 a pair of Franklin's Ground Squirrels brought from Illinois escaped in New Jersey. Reportedly they established themselves in sandy fields and spread northward in the state.

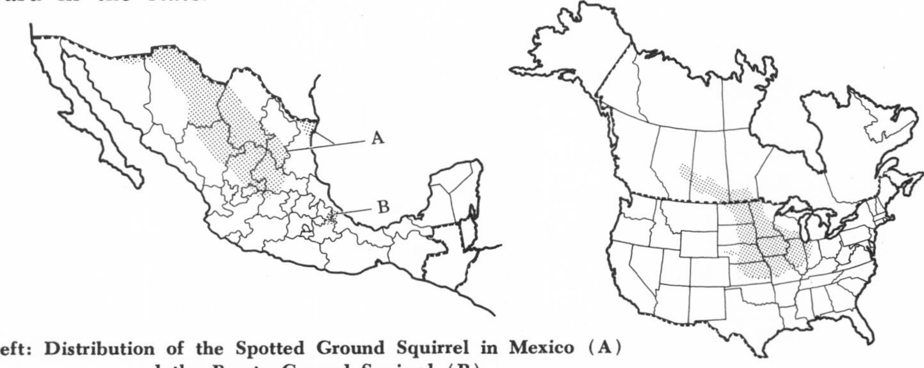

Left: Distribution of the Spotted Ground Squirrel in Mexico (A) and the Perote Ground Squirrel (B)
Right: Distribution of the Franklin's Ground Squirrel

Habits. Borderlands between woods and prairies are the preferred habitats of Franklin's Ground Squirrels. Seldom do they live in forested areas or on open prairies. Seton reports the gray gopher's occurrence to be restricted to the edges of woods in the prairie provinces of Canada; in southern Manitoba he found this squirrel living in aspen groves on the prairie as well as stream and marshland borders and along wooded lake ridges. Schwartz states that in Missouri these squirrels prefer "the marginal habitat afforded by fence rows, wooded banks, gullies and little-grazed sod."

Cover is important; Franklin's Ground Squirrels often live where the vegetation is so high they cannot see over it, so dense they cannot see through it for more than a few feet.

Franklin's Ground Squirrels usually live together in small colonies of ten to twelve animals. Often the colony moves its location. Individual squirrels occupy home ranges that are about a hundred yards in diameter.

A brushy bank or draw is the usual burrow site, but "in times of abundance," Schwartz notes, Franklin's Ground Squirrel "may establish its home around farmyards where it utilizes the shelter of shrubbery, trash piles, and buildings." Frank E. Wood writes of the species as an agricultural pest:

> As their burrows each have several openings and these are conspicuously marked by the dirt thrown out, a colony becomes a great nuisance in a hay or grainfield. The conspicuousness of these burrows and of the animals themselves has aroused the animosity of the farmers and hastened the destruction of the gophers.

And he mentions the squirrel's fondness for refuse heaps:

> They avoid closely cropped pastures, well-kept cemeteries, lawns and similar places where the striped gopher is especially abundant, yet even in such localities I have found them congregated under a heap of compost.

Of the burrowing habits of Franklin's Ground Squirrels in Illinois Robert Kennicott writes:

> It is fond of digging long burrows in the banks of ditches, and several times I have seen it living in steep river banks, as well as under small wooden culverts in roads. It is not so shy as the striped spermophile, and takes up its residence quite near dwellings.

Sometimes its home is the abandoned burrow of a pocket gopher.

In Manitoba male Franklin's Ground Squirrels emerge from hibernation about the third week in April. Sowls believes the squirrels hibernate in groups; after emergence from hibernation many squirrels appeared near a few dens. Succulent roots, cherry stones, green nettle shoots and grasses are eaten. The males battle among themselves—nearly every male bears a hairless spot or laceration on its rump as a result of these springtime skirmishes. When the females emerge from hibernation, about a week after the males, mating takes place, usually after a violent chase ending in a burrow. Litter size varies from four to eleven, but more commonly a female has eight or nine baby ground squirrels. The inch-long, naked, reddish pink babies develop rapidly. Short hairs cover their bodies by the ninth day, and the little squirrels are fully haired at sixteen days. Their eyes open at eighteen to twenty days. The babies explore their nest and make whistling noises. The young squirrels are weaned by early July and by mid-September have almost attained adult size.

Only about ten per cent of a Franklin's Ground Squirrel's life is spent above ground. Sunlight, wind and temperature affect its daily activity in spring and summer; cool, cloudy weather or a windy day restricts the squirrel to its burrow.

Examining the stomachs of 178 Franklin's Ground Squirrels, Sowls found most of the contents to be plant material—sow thistle, beach pea, chokecherry, red-berried elder berries, white clover, and roots and tubers, toads, grasshoppers, beetles, ants and other insects are included in its diet. Of the ground squirrel's carnivorous habits Howell writes:

> Individuals of this species have been known to kill and eat wild mice and a young rabbit, to rob a meadowlark's nest, to kill a wood pewee, to capture small chickens, and to eat hen's eggs.

And Hall adds:

Whenever I see one of these ground squirrels there is something about its slenderness and movements that suggests to me the Long-tailed Weasel, which is strictly carnivorous.

Franklin's Ground Squirrel has a reputation as destroyer of pheasant and duck nests. Sowls studied a population of Franklin's Ground Squirrels that lived along the marsh edges, phragmites-covered islands and wooded lake ridges bordering the southern shore of Lake Manitoba and within the confines of the Delta Waterfowl Research Station. During three years of observation the squirrels destroyed 19 percent of the duck nests within their habitat, mostly nests of river ducks. Although good swimmers, the squirrels raided only nine of the more inaccessible nests of diving ducks. To eat a duck egg the squirrel grasps the egg under its body and uses its hind feet to throw the egg forward against its teeth several times until a small hole is made in the shell, which it then enlarges by biting away pieces of shell.

Scattered broods or stray ducklings are sometimes attacked and killed by Franklin's Ground Squirrels, but a brood that stays together is usually safe. The fact that almost 90 percent of a Franklin's Ground Squirrel's life is spent in its burrow, much of the time in hibernation, limits its role as predator as well as a prey species for carnivores, hawks and owls.

Heavy fat layers are accumulated during the summer. By late October adult Franklin's Ground Squirrels are ready to begin hibernation; soon the younger squirrels follow their example.

ROCK SQUIRRELS

Five species of Rock Squirrels (genus *Otospermophilus*) occur in western North America, from central Washington to Baja California, Mexico and east into Colorado and Texas. Rocky canyons and rocky mountain sides are habitats preferred by some species of this genus and from this preference has come their common name. But other species live in open, wooded clearings, even ranging out into meadows. Their altitudinal range extends from plains at near sea level up to elevations of more than 8,000 feet.

About eighteen inches in total length and one to two pounds in weight, Rock Squirrels resemble Gray Squirrels. They are more heavy-bodied and their grayish, brownish-gray or blackish coats, often sparsely furred, are mottled or dappled with light gray or whitish spots. Their tails, long and bushy, appear somewhat flattened. The generic name *Otospermophilus* (*otos* is from the Greek meaning "ear") refers to the relatively prominent ears of this group of ground squirrels.

Like all ground squirrels, Rock Squirrels run into holes in the ground when danger approaches. They are good climbers about rocks, stumps, brush and even in the branches of trees. Food, carried in internal cheek pouches to below-ground storage, is used during periods of bad weather. Rock Squirrels become fat in late summer and early fall, and in the colder parts of their range they hibernate.

Other names for these squirrels include canyon squirrel, gray squirrel and digger.

ROCK SQUIRREL (*Otospermophilus variegatus*)

Description. Black and white hairs, often mixed with buff, give a grizzled appearance to the upperparts of this squirrel. The hairs are blackish basally. Black fur covers the head and shoulders of some subspecies. Yellowish white tinges the underparts, and the paws are light buffy. The tail fur is mixed black or brown and buffy white.

Distribution. From central Nevada, northern Utah and Colorado this species ranges into southeastern California, the Panhandle region, through western Texas and south into central Mexico.

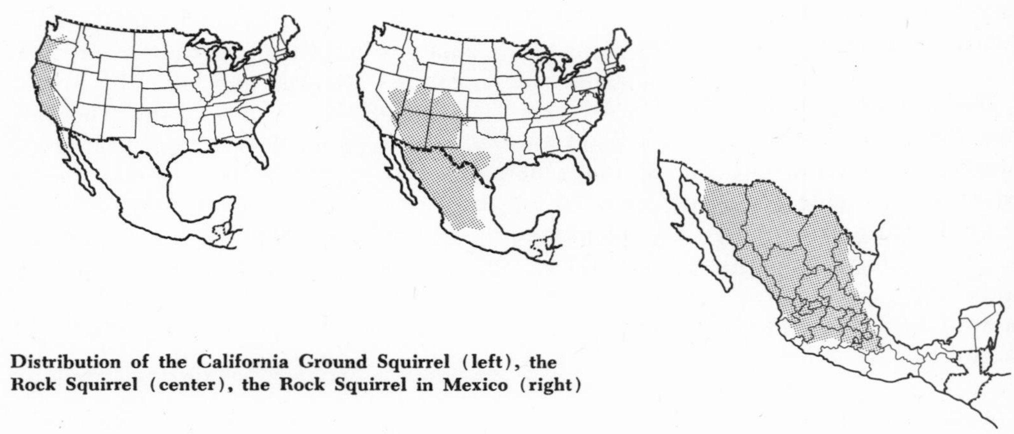

Distribution of the California Ground Squirrel (left), the Rock Squirrel (center), the Rock Squirrel in Mexico (right)

Habits. This species decidedly prefers rocky habitats. Its burrows are usually dug beneath a large boulder, or sometimes in thickets. In Mexico Rock Squirrels make their homes along stone walls or irrigation ditches. Unlike most ground squirrels, Rock Squirrels climb trees for food, and so occasionally find it convenient to live in hollow trees, fifteen or twenty feet above the ground.

Acorns, pine nuts and walnuts are eaten by Rock Squirrels; so are grasshoppers, crickets, caterpillars and other insects. Seeds of mesquite, cactus, saltbush, agave, wild gourd, wild fig and fruits and berries, either wild or cultivated (for Rock Squirrels are often considered agricultural pests), are eaten. Orchards of apricots, pears, peaches, plums, cherries and apples are sometimes raided, and such crops as melons, squash, peas, beans, corn and all kinds of grain are pilfered. Rock Squirrels climb juniper trees to harvest berries and climb up into mesquite trees to eat green buds or beans.

Rock Squirrels are food storers. During winter they depend on their underground caches of food stored under boulders or in hollow trees. In cold weather the squirrels venture above ground only on mild sunny days. Perhaps in the northern parts of their range Rock Squirrels hibernate. In hot, arid regions they may aestivate. But over much of their range Rock Squirrels remain active year-round.

CALIFORNIA GROUND SQUIRREL (*Otospermophilus beecheyi*)

Description. Gray, light brown and dusky fur are mixed to give the California Ground Squirrel's upperparts a mottled appearance. A band of slightly darker fur, flecked with light gray, extends from the head over the middle of the back. Gray fur forms a cape over sides of head and shoulders. This gray cape may have a protective function, breaking up the animal's body outline and making it more difficult for a predator to spot. Light buff or grayish yellow fur covers the undersides. Whitish fur rings the eyes and perhaps protects the eyes from too intense sunlight. Black fur edges the outer rims of the ears. The tail, five to seven inches long and more than half the length of head and body, is covered with mixed yellowish gray and black hairs and is lighter on the underside. California Ground Squirrels, measuring between sixteen and nineteen inches total length, are often slightly smaller than their Rock Squirrel cousins.

Distribution. From central Washington through western Oregon, California and into the northern part of Baja California this squirrel is found in plains, small meadows, tree-covered hillsides, rocky outcrops and granite taluses. The northern extent of its range was once bounded by the Columbia River, but the California Ground Squirrel has crossed the river into south-central Washington.

In California this ground squirrel has thrived with the agricultural use of land in the valley and foothill regions west of the Sierra Nevada. An increased food supply as well as reduced numbers of natural enemies contribute to large populations of California Ground Squirrels. Robert T. Orr writes:

> Today the artificial control of ground squirrels is a necessity in many farming areas because of the upset in nature's balance that inevitably accompanies advancing civilization.

However, Jean M. Linsdale reports that artificial reduction of a California Ground Squirrel population is generally followed by an increased reproductive rate among the surviving squirrels, a kind of compensatory gain. Where land is no longer used for grazing stock or growing crops the squirrels decrease in number or disappear.

Named for Frederick William Beechey, who explored much of Northern California in 1826–28 as captain of His Majesty's Ship *Blossom,* this squirrel is sometimes called the Beechey Ground Squirrel.

THE WAYS OF A CALIFORNIA GROUND SQUIRREL

In an open grassland area of central California lived a population of California Ground Squirrels. Their habitat was one of scattered trees and bushes, sparse low grass and dry, loose soil. The colony spread over a large area, with as many as seven squirrels living in each acre-sized plot. The ground squirrels' burrows were located along fences, under logs, in thickets of vegetation and in earth banks. Hillsides or low earth banks were preferred sites because the burrows could be excavated horizontally, but some of the burrows had to be dug down vertically for several feet to assure protection. Many of the deep burrows had been used by generations of ground squirrels.

CALIFORNIA GROUND SQUIRREL
(*Otospermophilus beecheyi*)

The burrows, four or five inches in diameter, varied in length from five to as much as thirty-four feet. Most of the tunnels were three or four feet below the ground surface. In a Fresno County, California, chalk pit a California Ground Squirrel's burrow that was excavated was twenty-eight feet below ground. Some of the burrows were short tunnels, many of the others were branched and had two or more openings. Some burrows had single squirrel occupants, others were colonial homes of several squirrels. A record colonial burrow was excavated in California's San Luis Obispo County. The home of six females and five males, its tunnels totalled 741 feet in length and had thirty-three openings.

Although the squirrels in the colony did most of their tunnel excavation in spring, digging and burrow improvement was a continuous process. The squirrels brought soil to the surface. Food stores, nesting material and feces were mixed with the soil. Their tunnels aerated the soil.

During the breeding season, from February to April, the ground squirrels were active above ground for long periods each day. The males, always aggressive and territorial, moved into short, shallow, simple burrows on the colony's margin. Nest burrows, occupied by females and their young, were complicated by numerous turns and blind passages. At this time of year the colonial burrows served mostly as places of refuge for both males and females and, later, for young ground squirrels.

A female California Ground Squirrel had just found a new burrow. While most of the squirrels were feeding, sunning, dust-bathing and grooming (for the squirrels were molting and new fur was growing in), the female ground squirrel was busy gathering long grass stems. She folded them many times with her front feet, then stuffed the cut stems crosswise in her mouth and made several trips into her burrow. She was building a nest for her soon-to-be-born litter and she worked hard to renovate the old burrow she had chosen for a nursery. After about a month-long gestation the female gave birth to seven babies. Litters in the ground squirrel colony varied from three to as many as fifteen babies, but her litter was average in number.

At five weeks the babies' eyes opened and by the eighth week they were ready to come out of the burrow for the first time. The young played and fed near the burrow entrance. Although they would stay with their mother for almost two more months the young squirrels received little attention from the female ground squirrel. Less wary than the older squirrels in the colony, the rapidly growing little squirrels were easy prey for such natural enemies as rattlesnakes, golden eagle, hawks, weasel, raccoon, badger, gray fox, bobcat and coyote. The surviving youngsters would be full grown by November, at seven to eight months of age.

Young California Ground Squirrels were active all over the colony during the summer. Adult ground squirrels were seen only at times, and by late summer only a few adults came above ground, for their period of dormancy had begun.

Responding to climatic conditions in their habitat California Ground Squirrels may become dormant in midsummer and again in midwinter. Linsdale writes of California Ground Squirrel activity:

> Adults may spend as many as eight months dormant in their burrows each year. Even in months when they are active, they are on the surface not more than a quarter of the time. This short active period is used for food gathering, den and nest building, establishing a social place, and absorbing sunshine. Accord-

> ing to our estimate made in July, in a 13- or 14-hour squirrel day from one to five hours are spent outside the burrow, foraging occupying at least half this time.

Linsdale believes the squirrels may aestivate after taking on fat but without storing food, and emerge in the fall to store food. By this time the active young squirrels have become fat, too. The squirrel colony then goes underground for a second period of inactivity, with stores of accumulated seeds and nuts.

California Ground Squirrels living in open forests or talus slopes of the higher Sierra Nevada, where there is snow for a number of months, remain dormant from October or November into March or April. Male ground squirrels are first to emerge from hibernation.

Throughout the colony in which the female and her young lived the life cycles of the ground squirrels were determined by their burrow sites. Occupants of burrows with open, sunny exposures remained active throughout the winter. But squirrels that lived in burrows on cold, shaded slopes tended to become dormant, at least during the cold, wet months of December and January. Many of the adult ground squirrels were inactive from June until March—a stretch of time that included the extremes of hot-dry and wet-cold conditions.

California Ground Squirrels feed on a large variety of seeds, fruits, acorns, roots and bulbs. Seeds of red brome, ripgut grass, barley, oats, tumbleweed, miner's lettuce, shepherd's purse, wild radish, poorman's weather glass and bur clover are eaten. Gooseberry, blackberry, peach, pear, green berries of manzanita and watermelon are favorite fruits. Valley oak, blue oak, coast live oak and California black oak supply abundant acorns in the fall. Mushrooms and such insects as grasshoppers, crickets, beetles and caterpillars are included in their food list. As for amounts eaten, Linsdale believes about two ounces of food make a meal and that the stomach is filled twice a day.

The California Ground Squirrel's fondness for quail eggs serves to limit quail populations where the squirrels are numerous. A squirrel that robbed a nest containing six quail eggs left shell fragments, tell-tale evidence, around its burrow entrance. Ground-nesting birds are not the only ones to have their eggs destroyed and their nestlings molested by California Ground Squirrels. Climbing among the branches of a tree the ground squirrels sometimes happen upon nests. One observer reports seeing a squirrel in a blue oak raid a mourning dove's nest. The squirrel consumed one egg and carried the other away.

Runways, or beaten tracks through the grass, are made and used by California Ground Squirrels when they travel from hole to hole or to favorite foraging places, especially where populations are heavy. Some runways interconnect to form a network, others simply extend from burrow to feeding area. Often the three-inch-wide runways go directly over small objects such as stones and sticks.

The female ground squirrel travelled along such a runway to the top of an embankment where tar weed flourished. To reach the seeds of the plant she pulled down stalks and even climbed up among the intertwined stems of the larger tar weeds. As the squirrel fed, seeds scattered to the ground. Three California quail, their top-knots bobbing, fed on the seeds. They followed along the slope below the squirrel, eagerly snatching the seeds.

Live-trapping and marking of California Ground Squirrels as well as field observations indicate that most squirrels are active within a hundred feet of their own burrows; rarely does a squirrel go beyond a 150-yard radius of its burrow. In areas of tall grass watching for enemies is difficult and squirrels seldom stray far from their burrows. On sparsely covered ground squirrels repeatedly sit up full length, with backs straight, to look about. Young ground squirrels move about more freely, some even moving to other areas.

Within its home range sunning is a ground squirrel's favorite pastime. Linsdale describes California Ground Squirrels sunning at the entrance of a burrow:

> In one favorite sun-bathing position the squirrel lies with its belly on the mound, elbows on the ground with forearms extended and head raised, or with forefeet on the ground holding the chest off the ground, and the hind feet under the body and flat on the ground. Sometimes the forelegs or both pairs of legs are extended and the chin is placed on the ground, the body in a prone position.

California Ground Squirrels sometimes climb trees for sunning, food or to reach a tree hole. Usually, at any hint of danger, the squirrel hastily leaves the tree and scurries to its burrow. Of a climbing ground squirrel Linsdale writes:

> One squirrel made repeated use of a large, dead oak within its home area on a canyon side. Once . . . the animal climbed 15 feet up the trunk of this tree, ran 30 feet horizontally along its only limb, then ran down a sloping branch and disappeared into the grass. It appeared perfectly at home.

Lloyd G. Ingles observed a California Ground Squirrel:

> Fifteen minutes after sunrise . . . emerge from a hole 20 feet up in a living tree where it had spent the night. It sunned itself several minutes then climbed far out on a branch where it prevented two gray squirrels from using the tree as an arboreal route. Later, using an arboreal route, it crossed the creek and came to the ground to feed.

Another arboreal ground squirrel was seen to climb a buckeye tree, nip off several nuts and run down to the ground. The squirrel removed the buckeyes' outer coverings and carried the nuts into its burrow. Almond and prune growers of California's Sacramento Valley are unhappily aware of the California Ground Squirrel's tree-climbing habits. The green fruits of the almond trees are relished by the squirrels.

When it emerges from its burrow the California Ground Squirrel is careful to stop and freeze for about three minutes. Then it runs off about its business, stopping every few feet with head up and body frozen. The squirrel either runs in a series of long, low bounds, fore- then hind feet striking the ground together, or it trots through runways or tall grass, with head down, tail straight out behind. If frightened, a squirrel often makes long leaps and emits its sharp, metallic alarm cry several times in rapid succession. The squirrel may pause near its burrow and *clink* at intervals, or it may drop down into its tunnel system.

Of all its enemies man, with his traps, guns, fumigants and poisons, has been the most relentless. California Ground Squirrels are persecuted not only for their threat to agricultural crops but because they have sometimes been disease carriers. Many squirrels have been eradicated in efforts to control outbreaks of such diseases as bubonic plague and tularemia. But where their homes are far from grain fields and orchards wary California Ground Squirrels may live as long as six years. Three or four years is probably an average life span for the species, although captive squirrels have lived for ten years.

BAJA CALIFORNIA ROCK SQUIRREL (*Otospermophilus atricapillus*)

Description. This ground squirrel closely resembles the California Ground Squirrel, but its fur is darker, especially on the head and shoulders. It also has a slightly longer tail and a somewhat smaller skull.

Distribution. The range of the Baja California Rock Squirrel covers much of the south-central portion of Baja California. The California Ground Squirrel occurs in northern Baja California, but there is apparently an area of low country about forty miles wide where squirrels of this group are not found.

Distribution of the Baja California Rock Squirrel (left), the Ring-tailed Ground Squirrel (right: A), and the Lesser Tropical Ground Squirrel (right: B)

RING-TAILED GROUND SQUIRREL (*Otospermophilus annulatus*)

Description. The pelage of the Ring-tailed Ground Squirrel has a grizzled appearance. Black hairs are mixed with cinnamon buff on the upperparts, with more black on the head and back. Reddish buff-colored fur covers the sides of head, neck, shoulders and forelimbs. The narrow tail, as long as the head and body, is marked by about fifteen blackish rings, or annulations.

Distribution. Ring-tailed Ground Squirrels occur along the Pacific coast of Mexico in the states of Nayarit, Jalisco, Colima, Michoacán and Guerrero. Tropical deciduous forests are home to this squirrel.

Audubon painted the Ring-tailed Ground Squirrel from a specimen ". . . obtained on the Western prairies, we believe on the east of the Mississippi river," about 1851. This same original specimen, from which the species was described, was subsequently considered to be an African squirrel. It was not until 1877 that Joel Asaph Allen restored the squirrel to its rightful place among North American squirrels, at the same time noting some of its resemblances to the tree squirrels (*Sciurus*).

Of their observations of Ring-tailed Ground Squirrels in Colima, Mexico E. W. Nelson and E. A. Goldman write:

> We found their burrows on hillsides among rocks; and again in the sandy flats, along walls and hedges bordering cultivated fields; they are equally at home in the silent and gloomy shade of the densest groves of oil palms, with a burrow under a mass of fallen palm fronds or sheltered by the thorny growth of mesquite and acacia. . . . The nuts of the oil palm, mesquite beans, cactus seeds and the fleshy fronds of the pear-leaved cactus, wild figs, moho nuts and a variety of other seeds and fruits make up their varied bill of fare. . . . In going silently along the trails leading through the dense palm groves and thickets of other trees . . . they may be seen gliding silently from log to log or from one bunch of brush or similar shelter to another, now stopping a moment to dig for a seed or sitting up on their haunches to eat some morsel and then on again. . . . At the first alarm they scurry away into the first shelter. They carry their tails in a curve quite squirrel-like in character and their motions are more light and agile than those of most spermophiles.

LESSER TROPICAL GROUND SQUIRREL (*Otospermophilus adocetus*)

Description. This species is similar to the Ring-tailed Ground Squirrel, but it is smaller (thirteen to fifteen inches in total length), its fur is paler and less reddish and its tail is not ringed.

Distribution. Like the Ring-tailed Ground Squirrel this species is tropical. Its range, inland from the coast in Michoacán and Guerrero, Mexico, overlaps that of its ring-tailed relative. The preferred habitat of the Lesser Tropical Ground Squirrel is more arid than that of the coastal species. According to Nelson and Goldman, these squirrels

> live among rocks along canyon sides, about stone walls and corrals near ranches, and sometimes their burrows are located in open ground at the base of a tree or bush. . . . dozens of these little animals were seen scampering along the trail ahead of us, sometimes playfully pursuing one another or sitting up to look about. As we drew near they ran to the stone walls and either sat on the top or took refuge in the crevices and with heads projecting from the holes watched us pass.

MOHAVE GROUND SQUIRREL AND ROUND-TAILED GROUND SQUIRREL

Grouped in the subgenus *Xerospermophilus* are two desert-dwelling ground squirrels easily identified by their tails. Both ground squirrels have very small ears and, like most desert mammals, pale fur. The ranges of the two ground squirrels meet but do not overlap along the eastern edge of the Mohave Desert.

MOHAVE GROUND SQUIRREL (*Spermophilus mohavensis*)

Description. About nine inches in total length, this squirrel has pinkish cinnamon fur above and white fur below. The white underside of its tail distinguishes the Mohave Ground Squirrel from other unspotted ground squirrels.

Distribution. Southern California's Mohave Desert is the home of this squirrel. Its preferred habitat is alkali sink areas and creosote bush scrub areas with sandy or partially gravelly soil. Sometimes it is found up in the Joshua tree belt.

Habits. Unlike the California Ground Squirrel and the White-tailed Antelope Ground Squirrel, which also occur in desert regions, the Mohave Ground Squirrel does not live in close colonies. William H. Burt writes of this species observed near Palmdale, California:

> In its more or less restricted range the Mohave ground squirrel usually is found on the lower desert, but penetrates the Joshua tree belt in certain places. Its preferred habitat in this part of its range seems to be where soil is sandy or of sand mixed with gravel, with a rather sparse growth of sagebrush. I never saw more than three or four to the mile, and often none for two or three miles. . . . They were distinctly less numerous than either *Citellus [Otospermophilus] beecheyi* or *Ammospermophilus* . . .

Scampering from bush to bush a Mohave Ground Squirrel carries its tail over its back, but its flag is not as conspicuous as that of the White-tailed Antelope Ground Squirrel, nor is it waved or twitched. Threat of danger sends a Mohave Ground Squirrel to its burrow, but almost immediately its head pops out again. Sometimes, when frightened, the squirrel crouches motionless on the ground, its protecting coat color blending with the sandy soil. Like most ground squirrels it keeps a wary lookout while feeding, sitting up picket-pin fashion, forefeet dangling, to survey its territory. Its call is a high-pitched, rasping *peep.*

Burt describes the behavior of a Mohave Ground Squirrel feeding in a patch of green vegetation, chiefly alfilaria:

> While feeding she would crawl along among the vegetation, bite off a green stem or leaf, hold it in her front feet, and eat it while sitting half erect. She would then crawl a bit farther and repeat the process. While searching for food . . . her tail was almost constantly moving slowly from side to side. When she chose to sit up her tail was brought against her back.

After feeding for nearly an hour this squirrel went into its burrow, then reappeared to climb over and explore the fallen trunks of a dead Joshua tree. Occasionally she climbed a foot or so up into small bushes to feed on green buds. The squirrel foraged over an area of about twenty-five yards, entering three different burrows, during the almost seven hours she was observed.

The Mohave Ground Squirrel is one of two species of ground squirrel living in the hot, dry, sparsely vegetated Mohave Desert; the other is the White-tailed Antelope Ground Squirrel. The two ground squirrel species are closely related; both are burrowers, diurnal and dependent upon the meager amounts of plant food available in the desert.

Although their food and habitat requirements are similar, the two species live together without competition, or are sympatric, because of differences in adaptation to their environment. While the Antelope Ground Squirrel has evolved a tolerance for high body temperatures and highly efficient kidneys that enable it to be active all year around in spite of extremes of desert heat and aridity, the Mohave Ground Squirrel's adaptation is its use of an underground retreat during the most rigorous times of year. From August to the end of February it remains in its burrow. This prolonged period, seven months, of aestivation and hibernation reduces the Mohave Ground Squirrel's competition with the more active and more abundant Antelope Ground Squirrel. During these months food is scarce and the desert is at its driest. Inactivity is the Mohave Ground Squirrel's adaptation to seasonal aridity.

Spring brings the desert vegetation to its peak. In March the Mohave Ground Squirrels emerge. Desert plants provide food enough for both ground squirrels. Mohave Ground Squirrels derive water from desert plants they eat; Antelope Ground Squirrels often travel considerable distances for drinks of water. Young Mohave Ground Squirrels are born in April. May, June and July are spent fattening on desert vegetation before they return underground in August.

Easily excavated in the sandy, gravelly soil, the burrow of a Mohave Ground Squirrel enters the ground at an angle of about thirty-five degrees. The soil is scattered about the burrow entrance so that no conspicuous mound is formed. Burt excavated a Mohave Ground Squirrel burrow system that had two openings about five feet apart. The openings were two and a half inches in diameter and led into a simple tunnel a foot below the ground surface that contained a single nest chamber.

The burrow in which a Mohave Ground Squirrel spends its dormant period is usually about eighteen feet long and goes as much as three feet below ground; a nest chamber is located at the tunnel's deepest point. When the burrow occupant retires underground in August it plugs the openings with soil.

Dormant from late summer to early spring, a torpid Mohave Ground Squirrel exhibits sharp drops in oxygen consumption and body temperature. At or just above environmental temperature its body temperature and oxygen consumption remain stable. To become torpid a squirrel may require as long as six hours. Breathing is suspended for long periods and heartbeat is reduced. Arousal takes less than an hour and is marked by increased breathing movements and oxygen consumption, more rapid heartbeat and shivering. Under laboratory conditions Bartholomew and Hudson discovered that Mohave Ground Squirrels alternate periods of torpor and wakefulness, three to five days of torpor at a stretch, during their months of dormancy.

ROUND-TAILED GROUND SQUIRREL (*Spermophilus tereticaudus*)

Description. This ground squirrel has two color phases: a pinkish cinnamon phase and a drab gray phase. In winter its fur is full and silky. Spring molt changes the pelage to sparse coarse hairs. Whitish fur covers sides of head and underparts. The Round-tailed Ground Squirrel has large black eyes, external ears that are mere rims and a tail, described by the species name *tereticaudus,* that appears round. Shy and secretive, this ten-inch-long desert ground squirrel is sometimes glimpsed sitting in the shade of a bush or fence post.

Distribution. From southern Nevada, southeastern California and western Arizona the Round-tailed Ground Squirrel ranges into the northeastern portion of Baja California and into Sonora, Mexico. It prefers to live in the lowest, hottest and driest desert regions, where wind-drifted sand forms mounds about the bases of mesquite, creosote or saltbush.

Distribution of the Mohave Ground Squirrel (A) and the Round-tailed Ground Squirrel (B)

Habits. In spring, when the desert is transformed after brief rains and succulent food is plentiful, Round-tailed Ground Squirrels are most active. After a long winter spent in their underground burrows the squirrels emerge to feed eagerly on leaves and buds of desert plants and soon grow round and fat. At other times of year seeds are eaten and stored. Seeds of plantain, goose foot, cactus fruit, blossoms of ocotillo and mesquite beans and leaves and insects are food for this species.

During March and April the young are born, from four to twelve in a litter. Later in the season second litters may be produced. But the squirrel population is constantly held in check by the always-hungry desert predators.

Round-tailed Ground Squirrel burrows are some two inches in diameter and five or six feet long. Usually the burrows are about a foot, but sometimes as much as three feet, below the ground surface. Often there is a grass nest in the burrow. Rapid digging with well-clawed forefeet followed by powerful backward kicks of densely haired broad hind feet excavates most Round-tailed Ground Squirrel burrows, but some squirrels move into kangaroo rat burrows.

A Round-tailed Ground Squirrel will run into *any* burrow for refuge, a useful safety measure. But frequently it reappears too soon and is snatched by a waiting predator. Badgers and coyotes often dig these squirrels out of their burrows, gopher snakes follow them underground and hawks and ravens take their tolls, especially of the unwary young.

Edmund C. Jaeger once observed a Round-tailed Ground Squirrel for a four-hour stretch and gives this account of the squirrel's daily activities:

> I saw him first as he emerged from a burrow. He evidently had fleas, for among the first things he did was to give himself a thorough going-over with both teeth and paws, scratching every part of the body that he could reach. Then he tidied himself by licking, particularly on the breast, belly, and forepaws, just as a cat does. Then for a moment he sat straight up on his haunches with tail up behind his back, the forelegs hanging relaxed over the chest and head turned slightly to one side, as the prairie dog does. It really surprised me to see how relatively inconspicuous he was during this time. The gray-buff color of his furry coat harmonized remarkably well with the sandy surroundings, especially when the sun was brilliant. For a minute or more he seemed to listen intently. Then several times he turned his head to look about. After that he resumed his former position on four feet and began to crawl along. For an hour or more he nosed about and fed on the succulent herbage which grew abundantly in scattered patches all about him. Twice during this time he assumed his alert position and gave an explosive, high-pitched, rasping note. When he ate he would sit up semi-erect on his haunches and, while holding the food in his forepaws, repeatedly nibble off small bits, moving his jaws very fast.
>
> Up to 2:30 P.M., when he scratched himself again and went down a nearby burrow, he had moved not more than 60 feet from the place where I first saw him. At 3:12 he came to the surface, but from an accessory tunnel, and began feeding again. Several times he varied his diet by climbing into bushes of desert tea for the small black seeds, first removing the husks or scales while holding the small "cones" in his forepaws.
>
> His only real excitement during the period I had him under observation was when a jack rabbit came loping along. The rabbit seemed unconcerned with the squirrel, but the ground squirrel, much excited, first sat upright on his haunches, then crouched low, and made a quick retreat into a kangaroo rat hole, giving vent to several sharp peeps on the way. He stayed underground this time fully thirty minutes. Several subsequent periods of watching round-tailed ground squirrels made me feel that on this day I had had quite a fair sample of the ordinary activities of these animals.

ANTELOPE GROUND SQUIRRELS

Five species of antelope ground squirrels occur in the arid plains, deserts and lower mountain slopes of western United States and northern Mexico. Most small burrowing desert rodents are nocturnal, but antelope ground squirrels, being ground squirrels, are diurnal. Characteristically, these lively squirrels bounce over the ground, carrying their small, flattened, white-backed tails over their backs. Of this habit Victor H. Cahalane writes:

> Scooting across the desert, dodging from creosote bush, to rock, to mesquite clump, the little antelope squirrel keeps its tail flattened protectively over its

> rump and back. As the tail twitches convulsively and rapidly, the exposed white undersurface reflects the sunlight. The dun-colored body blends into the sand and gravel. Like a ghost with the heebie-jeebies, the white tail seems to flee erratically all by itself. One is faintly reminded of the antelope and its going-away signal, which has given this ground squirrel its name.

Like the Golden-mantled Ground Squirrel the antelope ground squirrel is often called a chipmunk. But it is a small (six to nine inches in total length) and rather chunky striped ground squirrel. The lateral white stripes, one on each side, do not reach to the tip of its nose, as a chipmunk's stripes do. In true ground squirrel fashion antelope ground squirrels live in burrows they dig or take over, scamper over rocky slopes in arid canyons or scurry through bushy growth on sandy deserts. Throughout the year antelope ground squirrels remain active. Cold weather may keep them inactive in their burrows, but they do not truly hibernate.

The generic name *Ammospermophilus,* from the Greek, literally meaning "lover of sand and seeds," describes the habitat and diet preference of "Ammos."

HARRIS' ANTELOPE GROUND SQUIRREL (*Ammospermophilus harrisii*)

Description. This is the only species of antelope ground squirrel that does *not* have striking white fur on the underside of its tail. Two white stripes down its back mark a pinkish cinnamon coat that is grayer on neck and shoulders and darker bordering the white stripes. The fur becomes mouse-gray in winter. Underparts are whitish. The tail appears blackish above, its hairs broadly tipped with white; below, mixed black and white hairs give the underside a gray color.

The species was named for Edward Harris, who gave Audubon the specimen from which the original description was made. Audubon's son, John Woodhouse Audubon, painted Harris' Antelope Ground Squirrel with the California vole, a curious companion since the ranges of the two animals do not overlap.

Dr. Edgar A. Mearns, in *Mammals of the Mexican Boundary of the United States,* writes of Harris' Antelope Ground Squirrel:

> As one rides over the mesquite flats, it scurries from underfoot, carrying its tail straight up in the air, uttering explosive chipperings as it hurries to the nearest mesquite bush, under whose shade it is quite certain of finding numerous holes by which to make its escape; but it oftener stops and chirrups saucily, stamping with its forepaws.
>
> Its curiosity is so great that a few sharp chirrups with one's lips will often bring it to the entrance of its burrow, or it may run directly up to within a few feet of one. Then it stops, stamps, and jerks its tail, presently beating an equally precipitate retreat and diving into its burrow with a whistle; and it also utters metallic chirrups and chipperings suggestive of its impulsive nature. It very commonly sits up perfectly erect upon its hind feet, like the prairie-dog.

Distribution. South, central and northwestern Arizona, southwestern New Mexico and into Sonora, Mexico. In parts of its range the species is known also as Yuma Antelope Squirrel and Gray-tailed Antelope Ground Squirrel.

Distribution of the Harris' Antelope Ground Squirrel (left) and the White-tailed Antelope Ground Squirrel (right)

WHITE-TAILED ANTELOPE GROUND SQUIRREL
(*Ammospermophilus leucurus*)

Description. Brownish mixed with cinnamon above, two white stripes extending from shoulders to hips, a tail that is white below, narrowly banded with black, and has blackish hairs tipped with white on its upper surface, distinguish this species. In winter its fur becomes noticeably grayish. The specific name, a compound of Latin words, aptly refers to the white underside of the flattened tail.

Distribution. From southeastern Oregon and southwestern Idaho the range of this species extends south into Nevada, Utah, western Colorado, northern New Mexico and Arizona, southern California and to the southern tip of Baja California, which boasts three subspecies of the White-tailed Antelope Ground Squirrel.

THE WAYS OF A WHITE-TAILED ANTELOPE GROUND SQUIRREL

A nest hole in the gravelly earth opened beneath a creosote bush. It was the entrance to a burrow of a White-tailed Antelope Ground Squirrel, one of four that lived in an acre-sized area of rocky foothills on the edge of the valley. Rocks and gravel, carried down from the mountains by flash floods, spread in great fans over the lower slopes. No mound of excavated dirt marked the burrow entrance—the hole was so inconspicuous that when the squirrel scurried home in fright it seemed suddenly to disappear from view. Inside, the burrow went down almost straight for twelve inches before beginning its tortuous course among the roots of the desert shrubs that fortified the burrow walls against digging enemies. Only an industrious badger could dig successfully in the rocky soil, tear away the roots and perhaps find an antelope ground squirrel at home. Antelope ground squirrels themselves find digging difficult work. Sometimes they move into abandoned pack rat dens or take over kangaroo rat burrows. Whatever the origins of their burrows, the extensive network of underground tunnels always forms an effective trap for rain water in the

WHITE-TAILED ANTELOPE GROUND SQUIRRELS
(*Ammospermophilus leucurus*)

hard-packed desert soil. Their digging efforts and burrow maintenance result also in soil-mixing.

Deep in the burrow the female antelope ground squirrel had prepared a nest. Feathers and soft plant fibers and some fur from a rabbit, run over on the desert highway, made a soft nest lining. Late in April eight babies were born. By the time the young were covered with fuzzy hair and less than half their mother's size they came above ground. Their nervous mother seemed to pay little attention to them. While the young squirrels played near the burrow entrance their mother was busy gathering the black seeds of a yucca. In spite of the bayonnet-like leaves the squirrel climbed to the top of the yucca to get at the plant's plump, ripening pods. Cactus fruits, relished by antelope ground squirrels, could also be harvested with the same disregard for prickles and sharp spines. In addition to cacti and yuccas, seeds of

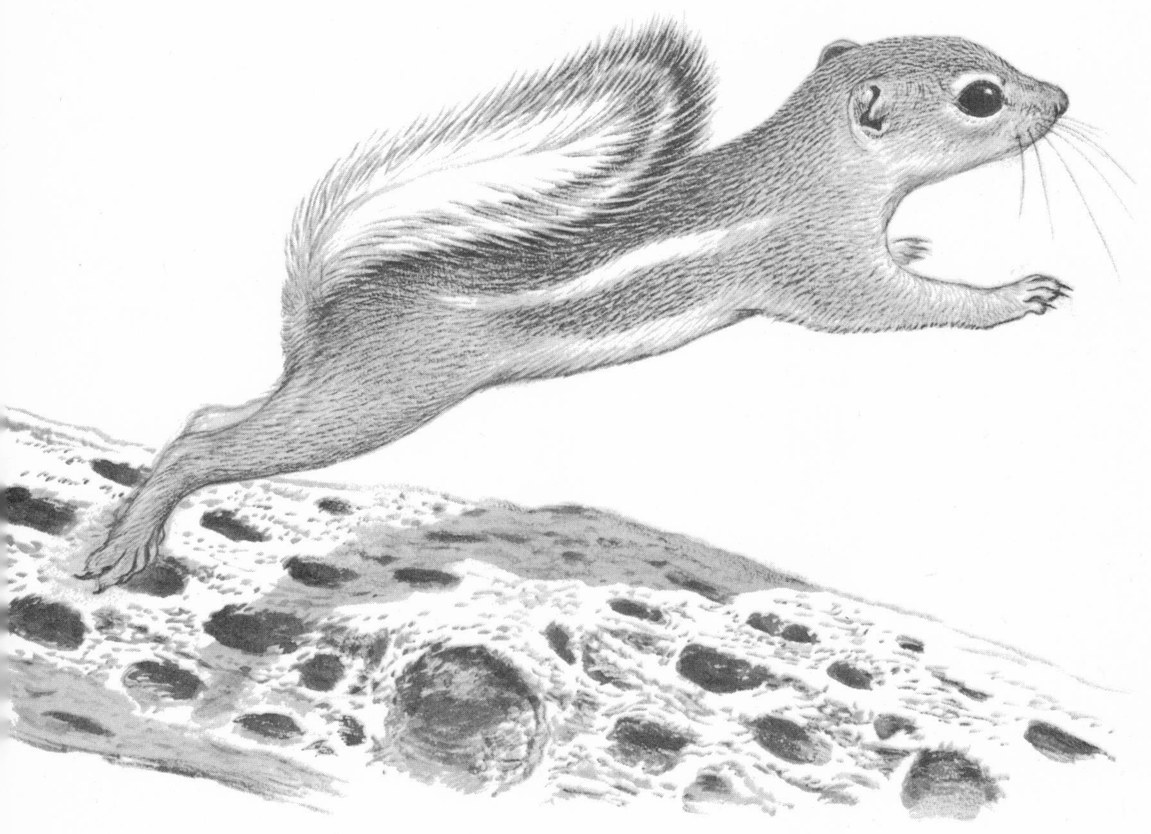

WHITE-TAILED ANTELOPE GROUND SQUIRREL
(*Ammospermophilus leucurus*)

ocotillo, sotol, palo verde, mesquite (Mearns once counted a total of forty-four mesquite beans in an antelope ground squirrel's capacious cheek pouches), various acacias, saltbush, sumac, tumbleweed, alfilaria, wild plum and piñon provide food for antelope ground squirrels. Jaeger describes White-tailed Antelope Ground Squirrels as "voracious feeders' 'and admits "being guilty of having several times, when in need of food, robbed the pantries . . . for the well-husked clean nuts" of *Simmondsia* or goat nut. Beetles, crickets, grasshoppers, grubs and sometimes a fly or ants are dietary supplements. Arthur B. Howell writes of captive antelope ground squirrels: "They demand meat in no uncertain terms, and when it is given . . . there must be enough to go around, or they fight viciously over it. Grubs, grasshoppers, chop bones, are pounced upon with avidity." Even where irrigated farm lands are nearby antelope ground squirrels are seldom guilty of poaching cultivated fruits. Perhaps they dislike the unfamiliar feel of damp soil and moist vegetation.

Unlike some desert-dwelling mammals antelope ground squirrels drink water when they can find it. Rains are infrequent and streams flow only at times in their desert habitat. Often the squirrels travel long distances for water. Jaeger writes:

> On hot afternoons I repeatedly watched these rodents quench their thirst by lapping water like a cat from a shallow drinking pan, and then afterward observed them go away to rest in the shade of the mesquite tree nearby. They sometimes contentedly squatted flat on their bellies with the inner face of the "hams" of their hind limbs flat against the sand, a position I have often seen taken by fat little dogs.

Desert temperatures often soar beyond tolerance of animal tissues. Mid-afternoon temperatures on sparsely covered desert floors may register as high as 180 degrees. Diurnal animals must seek sheltered places and wait out the heat. Small animals heat up rapidly under such conditions—although well-adapted to its habitat, the antelope ground squirrel is no exception. Now and then during a hot day it must get rid of body heat accumulations. By flattening itself against the soil in a shaded area or retreating into the relative coolness of its burrow, the antelope ground squirrel accomplishes reduction of body heat. Most desert rodents remain in their burrows during the day and are active in the desert's cool nighttime temperatures. But the antelope ground squirrel has evolved remarkable adaptations to the heat and aridity of desert life. Its metabolic rate remains constant when air temperatures range between 90 and 107 degrees F.; thus the species is said to have a broad thermal neutral zone. Since it can tolerate high body temperature its temperature can rise with that of the environment and it does not expend energy by panting or sweating to reduce body temperature. Only when its body temperature goes above 110 degrees does the antelope ground squirrel begin to show its discomfort. Then it unloads its heat accumulation by stretching out on the ground or going underground.

Antelope ground squirrels studied in the laboratory by George A. Bartholomew and Jack W. Hudson were found to lower their body temperature from above 107 degrees to about 100 degrees within three minutes after being transferred from an environmental temperature of 104 degrees to a temperature of 77 degrees. As a last resort drooling occurs. Under extreme heat stress the squirrel spreads saliva over its cheeks and head with its forepaws, attempting to cool itself by evaporation. In California's Mohave desert the White-tailed Antelope Ground Squirrel and its close relative the Mohave Ground Squirrel share the same habitat—an unusual instance of sympatry. But the Mohave Ground Squirrel solves the problems of heat and aridity in a different way.

At temperatures of 100 degrees an antelope ground squirrel loses three times as much water, through respiration and evaporation during a day, than it extracts from its food by oxidation. Yet it can survive as long as three to five weeks on a completely dry diet. In addition to moisture, which it converts from seed starch, an antelope ground squirrel nibbles green vegetation and the pulp of cactus stems and fruits as well as relishing insects and bits of flesh from animal carcasses. Like other desert rodents antelope ground squirrels excrete very concentrated urine. The renal papillae of their kidneys contain very long tubules for highly efficient water absorption.

Above ground the antelope ground squirrel was always on the alert. Her large black eyes were quick to sense the approach of any moving object. Often she paused while feeding or running along and looked up to be certain all was safe. Now and then she sat upright on her haunches, alertly listening. Jaeger notes the wary behavior of antelope ground squirrels:

> . . . they stop, spasmodically twitch the tail, look excitedly about, and turn the head a bit sideways in listening attitude. They soon bound away a short distance, assume the same alert attitude and violently vibrate the tail again. As they move toward a place of safety, they go from bush to bush, trying continually to keep behind them a screen of vegetation.

By wariness and caution the ground squirrel escaped her enemies—such neighbors as sidewinders and other rattlesnakes, weasels, bobcats, road runners, ravens and hawks.

Antelope ground squirrels have relatively large home ranges, and often the female travelled surprising distances from her home burrow. For emergency use, when danger threatened or she became too hot, she maintained several wayside or auxiliary burrows. It was toward one of these she raced, covering ground in leaps of eight or nine inches, when a prairie falcon's shadow flashed over the ground. She paused at the entrance, tail flicking, then plunged below. But the young antelope ground squirrels were unaware of the danger. The prairie falcon shot to the earth with great force, sunk its talons into the soft flesh of a young squirrel and bore it off to feed its own young. The other young ground squirrels fled in terror into the burrow. It was sometime before either they or their mother reappeared.

Antelope ground squirrels are late risers. Usually it is seven or eight o'clock, when the desert has warmed to their liking, before they emerge from their burrows and scurry over the ground to their feeding areas. A ground squirrel is always in motion, dashing from place to place, travelling hundreds of feet from its home burrow. After an active period of feeding the female antelope ground squirrel groomed her fur and stretched out on a rock for sunning. Suddenly she darted off to a retreat burrow to cool off. Although it was only a foot below the ground, and about twelve feet in length, the burrow was cool and welcome relief from the desert's mid-day heat. The burrow was supplied with some food stores but lacked the comfortable nest of the squirrel's home burrow.

During the afternoon the antelope ground squirrel family again became active. In winter this pattern would change—when the sun was out mid-day would be the time when the squirrels were most active. Cloudy weather, rain and sometimes snow would keep them underground, but antelope ground squirrels do not hibernate. Long, soft hair and warm underfur would replace the short summer pelage. The squirrels would depend on their underground food stores. But whenever it was sunny, no matter if snow covered the ground, the antelope ground squirrels would be out of their burrows.

TEXAS ANTELOPE GROUND SQUIRREL (*Ammospermophilus interpres*)

Description. Also known as the El Paso Ground Squirrel, this species reportedly differs from the White-tailed Antelope Ground Squirrel by being darker and more richly colored, and by having two black bands, instead of one, on the hairs of the tail. It has sometimes been regarded as a subspecies of *Ammospermophilus leucurus.*

Distribution. From south-central New Mexico and southwestern Texas into Coahuila, Mexico.

Distribution of the Texas Antelope Ground Squirrel, Espíritu Santo Island Antelope Ground Squirrel (A), Nelson's Antelope Ground Squirrel (B)

ESPÍRITU SANTO ISLAND ANTELOPE GROUND SQUIRREL (*Ammospermophilus insularis*)

Known only from Espíritu Santo Island, off the east coast of Baja California, this antelope ground squirrel is somewhat larger and has darker fur on flanks and legs than the White-tailed Antelope Ground Squirrel found on the Baja California mainland.

NELSON'S ANTELOPE GROUND SQUIRREL (*Ammospermophilus nelsoni*)

Description. Larger and paler than the White-tailed Antelope Ground Squirrel, this species has a buffy tinge (rather than grayish) in its pelage.

Distribution. San Joaquin Antelope Ground Squirrel is another common name for this species, found on the western side of the San Joaquin Valley, California, always on loamy soils.

Habits. In a study of Nelson's Antelope Ground Squirrel environment Albert C. Hawbecker concludes that the species' affinity for alluvial and loamy soils of the west-central and southern San Joaquin Valley and adjacent foothills was *not* related to the squirrel's digging habits. Numerous excavations of burrows indicated that although soils were well drained and diggable, this squirrel seldom lived in burrows that it dug. Instead, it made use of kangaroo-rat-excavated burrows. Hawbecker believes that this ground squirrel's range may be determined by the range of the kangaroo rat.

Climate affects the activity of Nelson's Antelope Ground Squirrel. As with other antelope ground squirrels mid-day is the period of above-ground activity in cold weather, while during warmer months early morning and late afternoon are times when the squirrels feed busily. Red-stemmed filaree and red brome are favored food plants. In moderate weather this species often stays out all day, spending only very short periods in its burrows. During the day, according to Hawbecker, a Nelson's Antelope Ground Squirrel travels over a circuit of hundreds of yards. An individual home range is about eleven acres.

Animals associated with this species include the horned lark, which feeds and nests on the ground. Probably the ground squirrels sometimes molest the nests as there are numerous reports of horned larks, as well as shrikes, chasing antelope ground squirrels. Horned larks and white-crowned sparrows are sentinels for the squirrels. If the birds are alarmed the squirrels take note. Hawbecker states it is impossible to stalk Nelson's Antelope Ground Squirrels when larks or sparrows are about. Occasionally living close by is the larger California Ground Squirrel (*Otospermophilus beecheyi*). Colonies of Nelson's Antelope Ground Squirrels and California Ground Squirrels appear able to live together, probably in much the same way as the White-tailed Antelope Ground Squirrel and Mohave Ground Squirrel do. Hawbecker writes that although one species may be dominant along the margins of the habitat, an established colony of antelope ground squirrels can withstand the larger squirrel's encroachment. In spite of the fact that a kangaroo rat once was seen fleeing from its burrow and an antelope ground squirrel intruder, and the evidence that many of the ground squirrels' burrows are excavated by kangaroo rats, the two animals usually avoid above-ground meeting. Kangaroo rats are nocturnal. Among Nelson's Antelope Ground Squirrel enemies Hawbecker notes red-tailed hawk, weasel, badger and coyote. If successful in avoiding too-close encounters with its natural enemies, an antelope ground squirrel may live as long as six years.

GOLDEN-MANTLED GROUND SQUIRREL

Of this squirrel Ernest Thompson Seton writes:

> More than any other of its group, it is a Rocky Mountain creature—the golden mantled climber of the golden ore-streaked rocks. The popular legend that there is gold wherever the Golden Chipmunk frisks, is at least worth smiling remembrance.

GOLDEN-MANTLED GROUND SQUIRREL (*Callospermophilus lateralis*)

Description. Often confused with the chipmunk, this handsome ground squirrel is larger (nine and a half to eleven inches in total length) and heavier bodied than its cousin chipmunk and wears a striking chestnut mantle. There are no stripes on the sides of its head, but whitish fur rings its eyes. On either side of its back are three longitudinal stripes—one white stripe bordered by black stripes. Grizzled grayish brown fur covers its back, while underneath the fur is yellowish gray or whitish. The soft, dense fur is shed during June and July, and by this time each year the rich, russet-colored hood appears worn and faded yellow. The three- to four-inch tail is brownish black above, edged with buffy, and yellow gray or reddish brown below. In winter the pelage is grayer and the mantle becomes dull.

Many common names have been given to this ground squirrel: calico squirrel, golden chipmunk, copperhead, big chipmunk and Say's chipmunk (the species was described in 1823 by Thomas Say, a member of Long's Rocky Mountain expedition). The Klamath Indians called this squirrel *Chil-las;* to the Paiutes it was known as *Wo-tah.* The generic name *Callospermophilus* is translated from Greek words meaning "beautiful seed-lover." The specific name *lateralis* refers to its conspicuous side stripes.

Distribution. The Golden-mantled Ground Squirrel occurs in the higher mountains of western North America, making its homes on mountain slopes and foothills and in open forested areas among rocks and fallen timber. Forests of yellow pine, limber pine, lodgepole pine, Engelmann spruce, Douglas fir and aspen, as well as mixed stands, support populations of these squirrels as long as the stand is not too dense. Open stands favor abundant ground-cover growth—a food supply for the squirrels. Direct sunlight penetrates open forest stands, at least for part of the day, and Golden-mantled Ground Squirrels are very fond of sunning on boulders or logs.

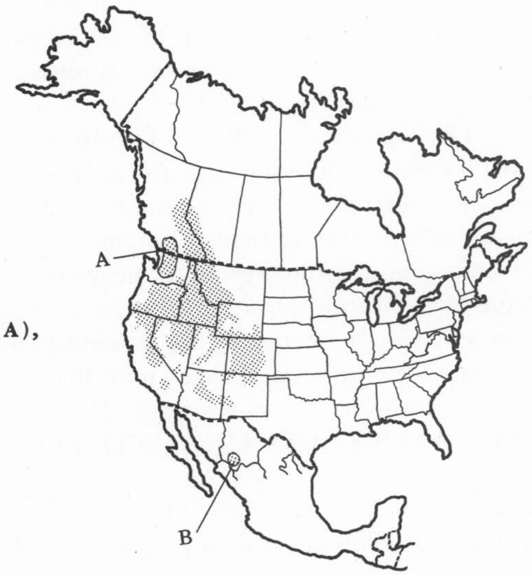

Distribution of the Golden-mantled Ground Squirrel, Cascades Golden-mantled Ground Squirrel (A), Sierra Madre Mantled Ground Squirrel (B)

A

B

Sometimes this species occurs in unexpected places. Robert T. Orr observed a Golden-mantled Ground Squirrel that lived in Farewell Bend State Park, on the Oregon side of the Snake River, an area of grass and scattered ash trees surrounded by desert. He writes:

> To our surprise a golden-mantled ground squirrel made its appearance and amused us for the next hour. It would race out on the open lawn, have confrontations with Brewer blackbirds then race back to a tree up which it would dash like a kitten. Half way up it would stop, whisk its tail and then leap to a picnic table. This sort of unusual golden-mantled ground squirrel behavior in an unusual golden-mantled ground squirrel habitat kept up as long as we remained.

In the Rocky Mountains the Golden-mantled Ground Squirrel's range extends from eastern British Columbia and western Alberta south to New Mexico and Arizona. It makes its home on most of the higher ranges in Nevada and in pine and fir forests of eastern Washington and Oregon. In California this squirrel lives in the high, inner north coast ranges, the Sierra Nevada and the San Bernardino Mountains.

THE WAYS OF A GOLDEN-MANTLED GROUND SQUIRREL

In a round globe of grass a Golden-mantled Ground Squirrel slept. His body was rolled into a heat-conserving ball, with nose tucked under body, forelimbs folded over chest, hind feet drawn underneath him and tail curled up over his head. In this position he exposed the least possible body surface to the cold air in his burrow. His winter of hibernation was about to end, but his breathing was still so slow that movement of his sides was barely discernible. The first sign of the squirrel's awakening was trembling. The trembling stopped and then, after nearly an hour, the squirrel began to shake his head. It was as though the front end awoke first—then the forelegs, body and rear legs. About twenty minutes later the squirrel stretched out, his eyes opened. He turned about in his nest, yawned and stretched each leg with toes spread wide apart.

The fat that the squirrel had taken to bed with him last October was all used up. Now, in April, the hungry squirrel must push his way out of the nest and down the tunnel to a storeroom. Here he found a few frozen kinnikinnik berries, hardly a satisfying meal. Winter freezing and thawing had caused the tunnel to cave in here and there along its length, but the squirrel pushed his way out into the snow and dazzling sunshine.

Snow was melting on the south-facing slope where the Golden-mantled Ground Squirrel made his home, and brown patches of earth already were touched with new green plant life. Cautiously the squirrel left his burrow to taste a few green shoots. He ate ravenously and retreated to his nest to resume his sleep.

About two weeks later, when there was little snow left on the south-facing slope, the squirrel emerged again. Caution marked his appearance. With his eyes just at ground level he stayed motionless in his burrow entrance and surveyed the slope and sky. Mountain sheep grazed on a high ridge across the valley, but they were not noticed by the squirrel. A deer walked through the meadow at the base of the slope but was not cause for alarm, nor were the Clark's nutcrackers that flew among the branches overhead, annoying the squirrel with their rattling calls. The squirrel decided it was safe to scurry along a worn squirrel trail to the top of a large boulder. For a long time he stayed there, motionless, absorbing the sun's mid-day warmth. The scream of a golden eagle was followed by its rapidly approaching shadow as it dropped downward. With a high-pitched chirp of alarm and a flick of his tail the squirrel plunged from the boulder and raced to the safety of his burrow.

During the spring the Golden-mantled Ground Squirrel lived for a time with a young female of his kind, sharing her burrow. He had located his mate by her scent. Golden-mantled Ground Squirrels have two sets of small glands just behind their shoulders. These are used to leave "messages" as they brush under branches and twigs. Some four weeks later, when five naked, blind and squirming baby squirrels were born, the squirrel moved out of the female's nest and built another burrow close by.

By this time people had come to spend the summer in the cabin on the south-facing mountain slope. The people enjoyed the squirrels and gave them food scraps from their hands. The young squirrels were more than one-fourth the size of the mother now, and they came above ground to scamper about and feed. Handouts of food supplemented a variety of seeds, berries, pine nuts, roots and bulbs. Yellow pine, Douglas fir, silver pine, oaks, serviceberry, gooseberry, kinnikinnik, Oregon grape, currant, thimbleberry, seeds of rose, lupine, bitterbush, shepherd's purse, milk-vetch, gilia and pentstemon all provide squirrel rations. Mushrooms are also eaten, and whenever the squirrels could capture an insect—grasshopper, caterpillar, beetle, ant or fly—it was eagerly devoured. Ernest P. Walker once observed a Golden-mantled Ground Squirrel that was especially fond of insects as food. The squirrel travelled some distance from its rocky home territory to a parking lot where it proceeded to examine the wheels of one car after another, relishing the squashed grass-hopper bodies it found on the tires.

As the Golden-mantled Ground Squirrels took food from their hands the people were amused at the way they stuffed their cheek pouches, then scampered off to their burrow storerooms, soon to reappear with empty cheek pouches and beg for more handouts. The holding capacity of their cheek pouches is large—two naturalists once counted 636 seeds from one Golden-mantled Ground Squirrel's cheek pouches.

The burrow of the mother squirrel had its entrance beneath a stump. Inside the tunnel narrowed to a two-inch diameter and sloped downward at an angle of about forty-five degrees. At a depth of nearly eight inches below the ground surface the tunnel levelled off. Approximately halfway along its length the tunnel forked. One passage led to the nest chamber, which was eight inches in diameter and four or five inches high. Dry leaves, shredded grass and bits of bark formed the nest. The other fork was a runway around the nest. Storerooms leading off the main tunnel contained seeds and acorns, some probably for use when bad weather kept the squirrel family in the burrow and some to feed on when the female would awaken from hibernation the next spring. Other side pockets were used for underground disposal of old nest material—instead of casting it outside the burrow where it might attract an enemy, such as weasel, badger or coyote. The horizontal tunnel, which extended almost ten feet, then slanted sharply upward to an emergency exit (or entrance), a hole camouflaged by kinnikinnik and a mat of leaves. Also living in the

GOLDEN-MANTLED GROUND SQUIRREL
(*Callospermophilus lateralis*)

burrow were fleas, mites, a few millipedes, a cricket and many ants. In spite of the time spent washing their faces, grooming their striking coats and taking dust baths, the squirrels always seemed to carry fleas that frequently caused them to pause and scratch.

Golden-mantled Ground Squirrels are somewhat territorial. The squirrel actively defended his territory, pursuing intruders of the same species for perhaps 100 feet from his burrow. Outside the limits of his territory, the Golden-mantled Ground Squirrel usually appeared indifferent to other individuals. Chases involving social dominance rather than defense of territory were less vigorous.

As summer gave way to autumn the Golden-mantled Ground Squirrel, as well as his mate and three surviving offspring (one was snatched up by a hawk, another pounced upon by a hungry bobcat), were growing noticeably fat. The squirrels added more leaves and grass to their nests in preparation for their approaching winter's sleep and carried many pouchfuls of seeds to their underground storerooms.

CASCADES GOLDEN-MANTLED GROUND SQUIRREL
(*Callospermophilus saturatus*)

Description. This squirrel resembles the Golden-mantled Ground Squirrel but it is somewhat larger (up to twelve inches in total length) and its mantle is a dull tawny color. The upper black stripe that borders the white lateral stripe on each side of its body is either much reduced or absent. The lower dark stripe on each side is shorter and not as well defined as it is in the Golden-mantled Ground Squirrel.

Distribution. From southern British Columbia through the Cascade Mountains of Washington. Some mammalogists regard this squirrel as a subspecies, now isolated, of *Callospermophilus lateralis*.

Golden-mantled Ground Squirrels often live side by side with the smaller chipmunks (*Eutamias*). Of the Cascades species Lloyd G. Ingles writes:

> The fact that two species in different genera with slightly different niches can live together is well illustrated by the Cascades golden-mantled ground squirrel and the yellow pine chipmunk of Washington. Both animals are abundant and are closely associated in the litter-strewn open places in the pine and fir forests on the eastern side of the Cascades. Both eat the same kind of food and appear to use the same places for their burrows. There seems to be little competition between them, probably because of the difference in niches.
>
> The Cascades golden-mantled ground squirrel gets nuts and seeds after they fall. The yellow pine chipmunk climbs bushes and high trees for the same food. The ground squirrel puts on thick layers of fat to provide it with energy during hibernation. The chipmunk draws from food stores in its winter burrow during torpid periods. Thus these two species live successfully sympatrically without serious competition.

SIERRA MADRE MANTLED GROUND SQUIRREL
(*Callospermophilus madrensis*)

Description. Smaller size, shorter tail and duller coloration distinguish this relative of the Golden-mantled Ground Squirrel. In fact this species wears only a faint mantle, its black stripes are short and poorly defined, but on each side of its body the white stripe extends nearly to base of the tail.

Distribution. Sierra Madre, Chihuáhua, Mexico. E. W. Nelson writes of this species:

> Very abundant in the pine woods about the base of Mohinora and reaches the extreme summit of the mountains. We saw them all along our route from above Guanacevi in Durango to Guadalupe y Calvo. Their range extends only a little below 7,000 feet and does not enter the pinyon belt.

CHIPMUNKS

American chipmunks are placed in two groups—the Eastern Chipmunk, *Tamias striatus,* and the western chipmunks, some sixteen species belonging to the genus *Eutamias.*

As the common names imply, *Tamias* is found in eastern North America while *Eutamias* occurs in western North America, for the most part west of the plains. Their ranges overlap in parts of southern Ontario, northern North Dakota, Minnesota, Wisconsin and Michigan. Gray coloration, narrower stripes and smaller size distinguish the western chipmunks from their rusty-colored, robust eastern relative.

The western chipmunks are more agile and often less shy than the Eastern Chipmunk. Many species of *Eutamias* are adept tree-climbers, climbing through underbrush, over down timber and even up into trees.

Tamias excavates a more extensive burrow system than does *Eutamias.* The hardy western species wait until snow and freezing temperatures have arrived before they disappear. Their hibernation retreat may be a nest in a conifer stump or log, or a burrow in the forest floor.

EASTERN CHIPMUNK (*Tamias striatus*)

Description. Five dark brown to blackish stripes, two gray or tawny stripes and two whitish stripes, and a bright chestnut rump are characteristic of the Eastern Chipmunk. Its body stripes end at the reddish rump. A dark stripe marks each eye, with a buffy stripe above and below it, and a reddish brown stripe on each cheek. The fur is grizzled rusty red to reddish brown above, whitish or buffy below. The tail is blackish above, rufous bordered with blackish underneath, and fringed with white or yellowish gray. Twice a year the chipmunk changes its fur coat; its winter coat is paler than its summer coat.

This lively squirrel, eight to twelve inches long, scurries with its flattened tail held straight out behind, or sometimes up. Scampering along old stone fences, pattering through meadow grass or scuttling in the underbrush, often its sharp "chuck-chuck-chuck" call is heard before the animal is seen. Numerous common names refer to its vocal habits, among them "chippie," "hackie" and "chipping squirrel." The scientific name *Tamias striatus* means literally "striped steward," alluding to the chip munk's tireless food-storing behavior.

As to how the chipmunk gots its stripes, Elsa G. Allen recounts this American Indian legend:

> The animals, led by the porcupine, held a council to determine whether there should be night or day in the world and after violent discussion, the chipmunk began to sing, "The light will come; we must have light," while the bear who favored night sang, "Night is best; we must have darkness." As the chipmunk sang the day began to dawn. Some of the animals became very angry and the bear ran after the chipmunk, nearly seizing him with his great paw. The chipmunk managed to escape into a hole, but wears still the black streaks made by the bear's paw.

Distribution. This species is found over much of eastern North America from Nova Scotia and the north side of the Gulf of St. Lawrence west to southern Manitoba, eastern North and South Dakota, extreme eastern Nebraska, Kansas and Oklahoma, and occurs in all the eastern states, including the northwest part of Florida.

Dry open woodland, forest borders, rock piles, stone walls and hedgerows and old outbuildings are ideal chipmunk habitats, offering plenty of shelter and an abundance of food. Tolerant of human beings, chipmunks sometimes frisk along walls and among shrubbery of suburban gardens. Merriam writes of chipmunk habitat preferences:

> He is partial to brush heaps, wood-piles, stone walls, rail fences, accumulations of old rubbish, and other places that afford him a pretty certain escape, and at the same time enable him to see what is transpiring outside. . . . he delights in these loosely sheltered hiding places, where he can whisk in and out at will, peep unobserved at passers-by, and dart back when prudence demands.

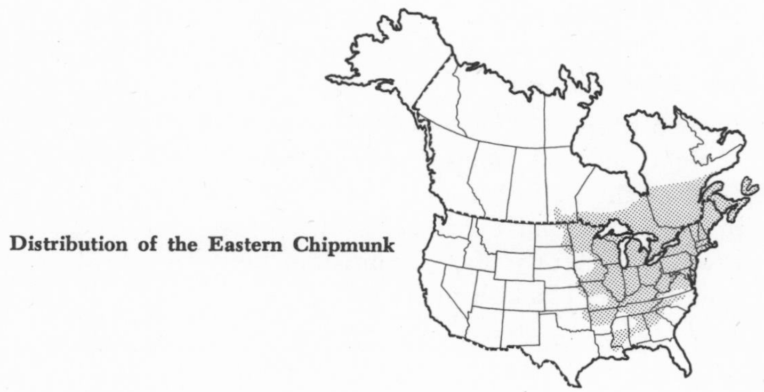

Distribution of the Eastern Chipmunk

Water is an important habitat factor for Eastern Chipmunks. Often they can be glimpsed drinking out of ponds and streams, and if they venture out in winter chipmunks lick snow to obtain moisture.

THE WAYS OF AN EASTERN CHIPMUNK

Deep in a burrow dug beneath an old stone wall a female chipmunk stirred from her winter nap. Finding the food supply hidden under her mattress depleted, she soon left the burrow. Male chipmunks had been out of their burrows for two or three weeks. Cahalane writes of chipmunk life in early spring:

> From February to the middle of March, the male chipmunk begins to go abroad regularly on pleasant days. Spring has wakened his emotions and he stops cautiously at the doorways of dens where the females are still staying. It is well for him to be circumspect . . . Female chipmunks are very fussy, and many a brash male has been beaten up and tossed out when he tried to rush things.

The woodland provided little food in early spring, but like all chipmunks the female had ground caches of nuts and seeds.

The year before the chipmunk had worked long and hard to excavate her underground home. She had dug down almost straight for nine or ten inches, then continued the two-inch-in-diameter tunnel on a twisting, sloping course that went nearly three feet below the ground surface. As she dug, loosened soil was kicked behind her in the tunnel. Now and then she paused, turned about and used nose and forepaws to push the soil outside the burrow opening. Chipmunks are always careful not to mark their burrow entrances by mounds of excavated soil. They may push the soil several yards away from the burrow or, as the female chipmunk did, continue to heap the soil just outside the burrow. After excavating a long tunnel from her nest up to the stone wall she went back to plug up the work-hole. Her inconspicuous front door, originally the burrow's exit, was protected by a large boulder in the base of the stone wall. There was no sign of recent digging to attract a passing predator, and soon a tangle of weeds would grow up to screen the small hole. Even this entrance tunnel could be plugged by a bit of earth when a fox or cat lurked near the stone wall. The chipmunk extended her burrow to a fourteen-foot tunnel. Several short side tunnels served as pantries, one was used as a toilet. The nest chamber, at the burrow's deepest part, was almost a foot in diameter, six or eight inches high, and filled with a bed of dried leaves and some grasses.

The chipmunk stuffed her cheek pouches with leaves and scurried off to her burrow. She was readying her nest for the birth of her young. Allen describes a mother chipmunk's burrows:

> . . . it was necessary to dig for eight feet, and a depth of about three feet was reached before any sign of a nest was found. The trend of the burrow was downward for two feet with slight turns, then horizontal for one and one-half feet. At two feet below the surface the burrow still proceeded downward for about 20 inches to a depth of 28 inches.

This tunnel made a sharp turn in the opposite direction for nine inches, turned again for twenty-two inches and then turned upward and into a large, one-foot-in-diameter nest cavity.

About thirty-two days after mating, the female chipmunk produced four tiny, red, hairless babies that squeaked and wriggled in the nest. By the fourth day their small ears stood out from their heads; at five days vibrissae appeared; at six days the baby chipmunks' lateral stripes began to show as unbroken hair follicles beneath their transparent skin and fine reddish brown hairs appeared on their snouts. Soon tiny incisor teeth pushed through their gums. At ten to twelve days the lateral stripes were conspicuous, the median stripe began to show, their eyes bulged beneath the lids. By the end of the fourth week short silky hair covered their bodies, and the babies were increasingly active in the nest. They crawled about and even attempted to sit up and wash their faces. Their coloration was that of adult chipmunks, including a bright reddish brown on their rumps. Except for their small, short-haired, rat-like tails the babies looked like miniature chipmunks. Thirty days after birth their eyes opened. During the second month the baby chipmunks grew to adult size.

EASTERN CHIPMUNK
(*Tamias striatus*)

It was mid-June when the four young chipmunks first emerged and foraged about their home burrow. At daybreak the chipmunks came out to gather breakfast. Berries of wintergreen, partridgeberry, bunchberry, dogwood, American yew, wild lily-of-the-valley, fruits and wild cherries are sought-after foods. Blackberries, blueberries, strawberries, elderberries and raspberries often stain chipmunks' faces and forepaws. Chipmunks are very fond of mushrooms. Seeds or fruits of box elder, maple, shadbush, viburnum, woodbine, red gum, star flower, buttercup, elm, prickly ash, wild buckwheat and ragweed are eaten. Perishable foods are seldom stored; even grains and nuts are seasoned before being cached away. In late summer and fall chipmunks are busy gathering hickory nuts, acorns, beechnuts, hazelnuts and walnuts. They seem preoccupied with nibbling, scurrying, searching and gnawing, and always they are watchful against enemies. Alan Devoe writes about the chipmunk:

> His universe is a smell of hickory nuts, a pungence of weed-seeds and of the dawn-damp earth to which his little muzzle is so close. . .

Chipmunks attracted to campsite or doorstep are omnivorous. Sometimes gardeners are provoked by chipmunks that dig up and eat planted seeds or excavate to gnaw on tulip, daffodil or other spring bulbs. But gardeners should remember that chipmunks benefit their gardens by eating various pests—cutworms, wireworms, millipedes, June bugs, beetles, slugs and snails. An empty snail shell that has a small, smooth-edged hole in one side of the coil usually indicates the inmate was eaten by a chipmunk. Chipmunks have been known to pounce on dragonflies and butterflies. Pinning the butterfly to the ground with its forepaws, the chipmunk sits upright, bites off the insect's wings and relishes its abdomen. Usually head, thorax and legs are discarded. One observer saw a chipmunk pursue a young bullfrog into shallow water. The captured frog kicked and struggled while being devoured headfirst by the chipmunk. Occasionally chipmunks raid birds' eggs or carry off the young of juncoes, bluebirds and robins. One chipmunk, in its haste to reach cover, is reported to have dropped a freshly killed meadow mouse. But such predaceous habits account for only a small part of chipmunk diets.

Of the good that chipmunks do Schwartz writes:

> Their life activities . . . contribute much to the successful functioning of a woodland habitat. Their tunneling helps aerate the soil and check rain and snow run-off; their food habits influence the growth of certain plant species and act as a partial check on insect populations; their bodies furnish food for many carnivorous animals and thus serve as an important link in the food chain of the wildlife community; and their body wastes and decomposed remains contribute to the fertility of the soil.

As the chipmunks fed, a high-pitched, soft conversation "cuck-cuck, cuck-cuck, cuck-cuck" was carried on. Even with cheek pouches full the chipmunks called back and forth, their tails twitching and jerking as they "chipped." This chipmunk call is often "sung" continuously for many minutes at the rate of 130 chips per minute. In spring and fall several chipmunks may form a chorus at a favorite site, sometimes attracting birds to their recitals. Now and then a chipmunk would scurry over the ground, its tail held horizontally. Or, on a tree-stump perch, a chipmunk would clasp its forepaws to its chest, peering with inquisitive bright eyes. Merriam writes:

> . . . he is very apt to advance toward the supposed enemy, betraying his excitement by a series of nervous starts and precipitous retreats, til finally, making a bold rush, he dashes by the object of his dread, and in another instant is peering out from a hole beneath the roots of a neighbouring tree.

Chipmunks are ever-alert to enemies, which include cats, foxes, coyotes, bobcats, weasels, rats, small falcon-type hawks, barred and screech owls and snakes. A loud "chip-chip" of alarm warns of a predator, but a chipmunk, suddenly frightened, utters a loud "chip-p-r-r-r-r-r" that ends in a trill, and dashes for cover with its tail straight in the air. Top speed for a chipmunk is about eleven feet per second. One of the four young chipmunks was overtaken in such a race for the stone wall. A stealthy fox pounced on him, quickly ending his short life. Chipmunk lives are also threatened by parasites—mites, fleas, ticks, bot fly larvae and roundworms.

Red Squirrels and Gray Squirrels often share the habitats of the Eastern Chipmunk. Although their feeding preferences are similar, the species usually get along. But Robert T. Hatt records a Red Squirrel that attacked a chipmunk which was gathering hazelnuts in the same tree. The Red Squirrel seized the chipmunk by the back of its neck, and both animals dropped to another limb. When the squirrel released its hold, the chipmunk fell lifeless to the ground.

In mid-July the mother chipmunk bred again. Observations conflict about a second chipmunk breeding season. Audubon and Bachman describe seeing half-grown young chipmunks in August. Seton states that a female chipmunk may produce several broods a year. Other naturalists maintain "second" litters are in fact "first" litters of females that failed to breed in spring. Allen concludes that "further observation is necessary to determine whether remating in the summer is the general rule or somewhat exceptional in this species." By this time the three remaining young chipmunks of the first litter were on their own. Each had started the excavation of its own burrow nearby.

Not strongly territorial, chipmunks usually coexist happily above ground. Allen writes:

> . . . territory in the sense of being a well-defined area within which a given animal stays and does not tolerate other individuals of the same species, does not obtain for the chipmunk. . . . wild chipmunks seem to enjoy communion with neighboring chipmunks, as evidenced by the chorus of their voices often so marked in the spring and fall. . . . The sense of ownership in the chipmunk seems stronger as regards its food supply than as regards its nesting ground or its mate. . . . though the animal may on occasion go one or two hundred yards away, it ordinarily spends its entire life within the compass of two or three acres.

John Burroughs also records evidence of chipmunk sociability:

> One March morning, after a light fall of snow, I saw where one had come up out of his hole, which was in the side of our path to the vineyard, and after

> a moment's survey of the surroundings, had started off on his travels. I followed the track to see where he had gone. He had passed through my woodpile, then under the beehives, then around the study and under some spruces, and along the slope to the hole of a friend of his, about 60 yards from his own. Apparently, he had gone in here, and then his friend had come forth with him, for there were two tracks leading from this doorway. I followed them to a third humble entrance, not far off, where the tracks were so numerous that I lost the trail. It was pleasing to see the evidence of their morning sociability written there upon the new snow.

But in a study of chipmunk ecology in central New York, R. W. Yerger reports that antagonism between animals of the same and opposite sex indicates that chipmunks *are* to some degree territorial, defending parts of their home ranges against intruding chipmunks. Home range is defined as the area over which an animal roams in its normal activities of feeding, mating and rearing its young; territory, within the home range, is the area defended by the animal against other individuals of the same species. Yerger writes:

> Defense of territory may be observed less frequently among chipmunks since these animals spend much of the day in an underground burrow, and thus do not sight intruders frequently. . . . territoriality may exist even where there is a broad overlapping of home ranges. . .

From four to fifteen chipmunks occupied an acre in this area of New York. Adult male chipmunks had home ranges of approximately 0.37 acre, while the females' home ranges averaged about 0.26 acre and young chipmunks' 0.18 acre. W. F. Blair, in northern Michigan, and W. H. Burt, in southern Michigan, found chipmunk home ranges to average a little more than two acres, with little size difference due to sex or age. Chipmunk home ranges overlap and usually are maintained by the same chipmunks for long periods of time, except when a male, or less commonly a female, moves to a new area. But the boundaries of a home range change in size and shape from week to week, season to season and year to year.

Chipmunks sometimes disappear, where they have been numerous, in mid- or late summer. Some observers have speculated that chipmunks are perhaps migratory. Merriam found chipmunks abundant in the Adirondacks in odd (or nut) years, and less numerous in the even years when the nut crop fails. Seton claims no one ever *observed* a migration of chipmunks, although Burroughs once wrote Seton of great numbers of chipmunks one summer at Roxbury, New York. Allen observed "a definite tendency for chipmunks to disappear during warm, dry spells, especially after the mid-summer litter is raised." And she adds: "the chipmunk is intolerant of very warm weather and wisely retires to his cool home below ground, provided he has sufficient stores of food to supply his needs." Aestivating for intervals of a few days to several weeks, according to Allen:

> . . . the chipmunk seems to be considerably more fossorial in its habits than the literature concerning it suggests. Its habit of storing food renders it rather independent for long periods of time whenever conditions above do not suit its fancy.

At least one mammalogist attributes this "summer disappearance" to the relative inactivity of the males and to the industry of the females foraging for their second broods, during which time there is little chipmunk calling to be heard.

Late summer and fall are times of scampering chipmunk activity. Each chipmunk works tirelessly to lay away its stores, which are usually far more than it will consume during the winter months. Because of its habit of caching large food stores beneath its nest and in its underground pantries, the chipmunk does not put on fat as do the ground squirrels. Many loads are carried away to storage in cheek pouches filled to capacity. Chipmunk literature records cheek-pouch loads of thirty-one large corn kernels, or two heaping tablespoonfuls, 145 grains of wheat, thirty-two beech nuts, sixty to seventy sunflower seeds, sixteen chinquapin nuts and thirteen prune stones. Audubon watched a chipmunk repeatedly carry four hickory nuts, two in one cheek pouch, one in the other and the fourth in its mouth. The chipmunk was careful to remove the sharp points on the ends of the nuts before stuffing them into its pouches.

The chipmunk does not store all its food in underground burrow pantries. Here and there under bushes, logs and rocks it digs small holes. To empty its cheek pouches the chipmunk pushes forward with its forepaws. Then the cache is carefully buried and leaf litter scratched over it for camouflage. Merriam observed a chipmunk store a half pint of buttercup seeds in a depression and cover over the cache. Another naturalist uncovered more than half a bushel of hickory nuts and acorns in four or five chambers of a chipmunk burrow he excavated. Burroughs experimented on the amounts of food a chipmunk stores away. The first day's offering, four quarts of hickory nuts and some corn, was eagerly carried away. Two quarts were removed for storage on the second day. The third day it rained and the chipmunk failed to appear. After carrying two quarts of chestnuts and one quart of hickory nuts set out on the fourth day, the chipmunk's hoarding instinct apparently was satisfied. Of chipmunk hoarding habits Allen writes:

> This gathering of food goes on regardless of the amount needed and usually continues until the supply is exhausted or until frost or other inclement weather conditions overtake the little steward and induce him to remain at home.

Throughout the southern states chipmunks are active above ground all winter, but in the north, sometime between late October and the end of November, chipmunks retire into their burrows. Occasionally several chipmunks share a nest for the winter. Allen observed chipmunks in winter under various conditions of temperature, darkness and dampness and concluded that chipmunks are true hibernators, often becoming torpid for a day or two at a time. Rolled up or sometimes curled on its side a hibernating chipmunk becomes cold and stiff. Now and then it awakens to nibble at the food stored under its bedding. Allen writes of chipmunk hibernation:

> In consideration of [the chipmunk's] ample food supply he has really no need for unbroken sleep nor the inhibition of his natural good appetite. Rather does it seem that he indulges it to the full, feeding now hourly, now daily as instinct demands, or occasionally food may not be taken for two or three days. At other times this capricious little mammal may be quite active in his burrow, and

> repair his nest or clean house by digging a new blind alley into which to pack the great mound of shells and shucks that continually accumulate about his nest.

Howell quotes C. C. Abbott, who dug out a chipmunk nest in New Jersey one November third to find:

> . . . four chipmunks very cozily fixed for winter, in a roomy compartment, and all of them thoroughly wide awake. Their store of provisions was in a smaller room or storehouse immediately adjoining, and consisted wholly of chestnuts and acorns; and the shells of such of these nuts as had been eaten were all pushed into one of the passages, so that there might be no litter mingled with the soft materials that lined the nest.

Another nest excavated by Abbott in March revealed two torpid chipmunks. Warming induced the chipmunks to become active. As to the effect of warmth on a hibernating chipmunk, Wirt Robinson records picking up a dormant chipmunk after workmen removed a boulder and exposed its burrow and nest of dried leaves:

> I found inside a chipmunk, tightly coiled up, eyes closed, cold to the touch and stiff and rigid. I moved it to another spot, placed it where the sun would strike it, and covered it with some dry leaves. Two hours later I returned and found it with its eyes open, but still stiff and unable to move. I put it in my overcoat pocket which I hung up in the warm building for an hour or so and forgot about the chipmunk. In putting on the coat later, I slipped my hand into my pocket and the chipmunk promptly bit me severely, its incisors passing through my fingernail. When I reached a suitable spot, I released it and it scampered off, now perfectly alert.

EASTERN CHIPMUNK
(*Tamias striatus*)

Probably few chipmunks live more than two or three years in the wild. Pet chipmunks, protected from their natural enemies, have lived as long as eight years. Allen describes a young chipmunk acquired when it was just out of the burrow. Fed at first on bread and milk, it later took hard foods. The chipmunk had the freedom of the Allen house and made a nest of leaves on the bottom shelf of a bookcase. She writes:

> Every evening at dusk it curled up in its nest to sleep, and before my two small children went to bed they picked it up in their hands and kissed it good-night without causing it to wake up.

WESTERN CHIPMUNKS (*Eutamias spp.*)

Description. This large group of chipmunks, some sixteen species in North America, is distinguished from the Eastern Chipmunk by its generally grayish appearance. Western chipmunks are smaller and their body forms are more slender; their tails are longer (almost half the total length); their ears are longer and somewhat more pointed at the tips; their stripes are finer and closer together. Two premolars on either side of the upper jaw are characteristic of *Eutamias.* The first upper premolar, a mere spike, distinguishes the western chipmunks from *Tamias,* which has only one upper premolar.

Eutamias, including the chipmunk of northern and eastern Asia *(E. sibiricus),* is not a strictly North American genus. It was not until 1823 that the naturalist Thomas Say described the first species, *Eutamias quadrivittatus,* from specimens collected on Major Long's expedition in eastern Colorado.

Five blackish and four whitish or buffy stripes mark this seven to ten and one-half inch-long chipmunk. The median dark stripe extends from occiput to base of tail. The outer lateral dark stripes are shorter and sometimes not distinct. The other stripes extend from shoulders to base of tail. Gray fur covers the head; the eyes are marked, above and below, by whitish stripes, and behind each ear there is in most species a conspicuous white spot. Grizzled brownish fur, soft, dense and rather short, covers the flanks. The tail is brownish black above, grayish underneath.

Distribution. From southern Yukon and southern Mackenzie, western chipmunks range south into central Baja California and Zacatecas, Mexico. Eastern Ontario, central Wisconsin, western South Dakota, eastern Colorado and western Texas form the eastern boundary of their range.

Western chipmunk habitats include spruce, fir, redwood and pine forests, brush-covered mountain slopes, sagebrush plains and rain forests of the Pacific northwest. Robert T. Orr writes:

> It would be a difficult feat indeed to travel very far through the mountainous portions of western United States without soon becoming acquainted with one or more kinds of chipmunks. They are almost as much a part of the scene as the coniferous forests in which they dwell. Not all chipmunks, of course, are restricted to the mountains or for that matter to tree growth of any sort. How-

ever it is with the pines and firs that we commonly associate these diminutive striped members of the squirrel family.

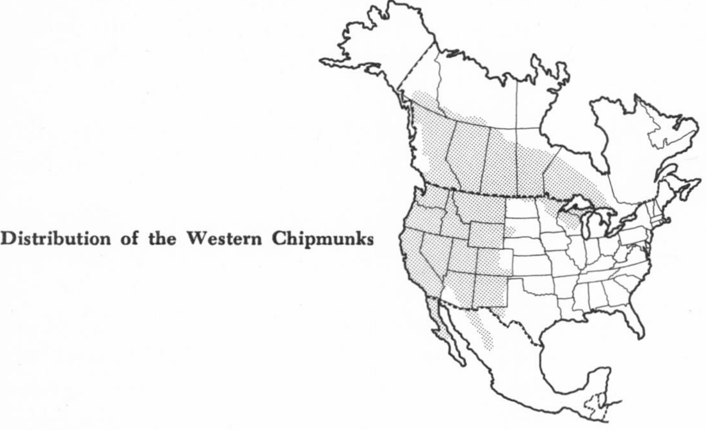

Distribution of the Western Chipmunks

Habits. Depending upon its habitat a western chipmunk's home may be a short burrow tunneled in soft, loose soil or beneath a rotting log, or a crevice in or under rocks. Although western chipmunks usually excavate their own burrows, they have been known to build outside nests of grass and leaves in shrubs or low trees and to move into abandoned pocket gopher and ground squirrel burrows or to occupy old woodpecker cavities in standing snags.

Excavation of a few western chipmunk burrows indicates they are usually small and round (about five inches in diameter), with a roof reinforced by roots of a stump or shrub. Other nests have been found inside rotten logs and even five to ten feet above ground in hollow tree stumps. Western chipmunks make their nests of shredded grass, frayed pieces of bark and moss.

Forest-dwelling western chipmunks are less shy than chipmunks that live in sagebrush areas. Often they become very tame and bold near picnic grounds and campsites. Many species are adept tree-climbers, and may go as high as seventy feet above the ground to gather food. But most arboreal travelling is from bush to bush through thickets or among rocks or fallen timber. Chipmunks have even been seen basking in the sunlit tops of bushes.

At intervals during the day western chipmunks scurry about looking for seeds, running across the forest floor, over logs and stumps and through clumps of small trees and brush. Often they call back and forth. The call may be a low-pitched resonant "cluck" emitted at regular and fairly long intervals. Or in alarm the chipmunk may give a rapid high-pitched chipping note, "whisk" or "whsst." Running to cover, a chipmunk gives a guttural jumble of notes. Of chipmunk calls Alden H. Miller writes:

> In a diurnal mammal . . . that makes frequent use of vocal utterances, . . . differences in call notes exist, which, as with birds, serve as releasers of instincts and so may coordinate activities and effect species recognition.

WESTERN CHIPMUNK
(*Eutamias quadrivittatus*)

Much of their time is spent collecting and storing nuts, berries and seeds. Western chipmunks include a large variety of foods in their diet. Acorns, hazelnuts, currants, blueberries, raspberries, salmon berries, huckleberries; such fruits as manzanita, buckbrush, buffaloberries, dogwood; seeds of grasses, smartweed, bindweed, creeping snowberry, cranesbill, thistle, lupine and asters; conifer seeds; birch seeds; cactus fruits; bulbs of camus and *Polygonum;* tender shoots and flowers of sagebrush, rabbitbrush, willow and dandelion; and mushrooms.

Perishable foods are eaten at once, but much of the food gathered during the day is cached for future use during occasional periods of waking or in early spring when snow still covers the ground. Cheek pouches filled to capacity, which may mean as many as 264 seeds of buckbrush or 1,650 seeds of wild cranberry, are emptied in scattered small caches in the forest floor, in rotten logs, under stones or in cliff crevices. Each chipmunk lays away its stores in so many scattered caches that some are never recovered. The buried seeds then sprout and grow. Chipmunks further assist in forest propagation by carrying conifer seeds into burned-over areas. *Eutamias* relishes many forest-damaging insects such as beetles, aphids, caterpillars and their pupae, and plays an important role in controlling forest insects.

Because of their arboreal habits western chipmunks sometimes molest nests. They have been reported pilfering wood pewee eggs. Cahalane writes of the hazards these little chipmunks face when they annoy larger birds:

> I once watched several birds, including a piñon jay, join in mobbing a chipmunk that was high in a pine tree. I was unable to determine whether it had gone there after eggs or bird flesh, but it seemed thoroughly unhappy about the consequences. When the coast was clear, it came down the trunk, head foremost, in a great hurry.

Dashing for cover with a shrill "tk, tk, tk," the chipmunk holds its tail straight up. Chipmunk enemies include snakes, hawks and most of the smaller carnivorous mammals. But weasels are probably their most dreaded enemy, sometimes pursuing a hapless chipmunk through the branches of a tree until it is exhausted, or even following the chipmunk into its nest.

Breeding season for western chipmunks may begin as early as late March. Some species produce two litters of two to seven babies in a season. At higher elevations in the Sierra Nevada and Rocky Mountains mating is delayed until May. The young are born a month later, and by July or early August the young chipmunks are out foraging on their own.

Most species go into dormancy during the winter. From mid-November until mid-March western chipmunks are rarely seen. Often they wait until the last minute to begin their winter sleep. Cahalane reports western chipmunks "wandering around in late November at eleven thousand feet elevation in northwestern Colorado when the snow was three feet deep and after the temperature had been 'way below zero." In January Alex Walker found a Townsend's Chipmunk, dormant and "curled up so cold and stiff that it could hardly be straightened," in its nest chamber five feet up in a hollow spruce stump. Late in winter A. W. Anthony dug up several fir stumps on an Oregon hill slope to find two or three chipmunks curled stiff and cold in hibernation, each in its own nest of shredded plant fiber. In the southern parts of its range *Eutamias* sometimes remains active for at least part of the time in winter.

According to Cahalane, there is a chipmunk species for nearly every habitat in the west:

> A pale little fellow lives on the brushy, arid plains. The tawny chap that has adopted the humid evergreen forests of the northwestern coastal belt has grown almost as big and as brightly colored as the eastern chipmunk. Some western species live in dark rocky canyons while others prefer the open slopes where the tall trunks of yellow pines stretch toward the sky like columns of a cathedral. . . . still other chipmunks [thrive] on heights where the stunted conifers struggle against the wind, cold and starvation of rocky, thin soil, and even beyond the zone of tree growth.

And H. E. Anthony writes:

> There is scarcely any peculiar ecological association in western North America which does not have its own peculiar form of Chipmunk, provided the animal can find food there. This accounts for the great number of known species and subspecies which, however they vary, are yet easily recognized as members of the genus *Eutamias.*

In California chipmunk species are segregated both zonally and ecologically and seldom are two species to be found in the same locality. But, writes Miller, ". . . because of interdigitation of habitats, one may find different species in close proximity, and even occasionally about the same trees or bushes."

Habitat and chipmunk size and coloration are correlated. The largest, dark-colored chipmunks are found in the humid coastal forests of southern British Columbia, Washington, Oregon and California. In arid plains and desert regions small size and pale coloration characterize chipmunk species.

Mammalogists have placed the various species of western chipmunks in five groups:

1 *ALPINUS* GROUP

ALPINE CHIPMUNK (*Eutamias alpinus*)

This small grayish chipmunk has weakly contrasted tawny side stripes on face and body. It is found in the central Sierra Nevada from elevations of 8000 feet to timberline at 11,000 to 12,000 feet, and lives on talus slopes as well as forest floor.

2 *MINIMUS* GROUP

LEAST CHIPMUNK (*Eutamias minimus*)

Bright coloration, ranging from yellowish to a rich grayish fulvous, and well-defined dark stripes characterize the many subspecies of Least Chipmunks. In general, chipmunks that inhabit sunlit open forests, where conifer twigs cast dark shadows, have well-defined light and dark stripes.

Wide-ranging both geographically and altitudinally, Least Chipmunks occur in western Canada, the northern Great Lakes states and throughout much of the Rocky Mountain region; they are "at home" in vast expanses of sagebrush and in coniferous forests.

3 *AMOENUS* GROUP

YELLOW-PINE CHIPMUNK (*Eutamias amoenus*)

The small- to medium-sized chipmunk is conspicuously striped and richly colored. Head and rump are grizzled gray-brown; sides of the body are bright reddish brown; the underside of the tail is yellowish. Yellow-pine Chipmunks are found from northern British Columbia through the Rocky Mountains in Wyoming and Idaho, in Washington, Oregon and northern California and along the eastern slope of the Sierra Nevada south to Mono County.

Often associated with the Cascades Golden-mantled Ground Squirrel, Yellow-pine Chipmunks climb conifers to get seeds, while the ground squirrel collects its seeds from the ground.

4 *TOWNSENDII* GROUP

TOWNSEND'S CHIPMUNK (*Eutamias townsendii*)

These large chipmunks have dark, reddish brown coloration. Their stripes are usually brownish rather than blackish. Townsend's Chipmunks that live in coastal forests are darker, illustrating a biological principle that mammal species in more humid areas are more heavily pigmented.

From southern British Columbia this species ranges south through western Washington and Oregon to central California along the coast and south into the central Sierra Nevada and extreme western Nevada in the vicinity of Lake Tahoe.

SONOMA CHIPMUNK (*Eutamias sonomae*)

Like the Townsend's Chipmunk this is a large, dark-colored species. Poorly defined body stripes conceal the Sonoma Chipmunk in its dense chaparral habitat where small twigs cast faint shadows.

Sonoma Chipmunks live in northwestern California, in chaparral and in open places in redwood and in yellow-pine forests.

MERRIAM'S CHIPMUNK (*Eutamias merriami*)

A conspicuously long tail (usually more than 80 percent of its head and body length), grayish coloration and indistinct stripes characterize this large chipmunk.

Merriam's Chipmunks occur in chaparral, in foothill vegetation and up into open coniferous forests in the southern half of California. Two subspecies are found in Baja California.

CLIFF CHIPMUNK (*Eutamias dorsalis*)

The Cliff Chipmunk is medium-sized and gray. An indistinct brownish black stripe runs down the middle of its back; the other stripes are nearly indistinguishable. E. Raymond Hall writes: "the median stripe is more pronounced than the lateral stripes but, at a little distance from the observer, the animals, particularly when in winter pelage, appear to be grayish without stripes."

Piñon pine-juniper areas, and rarely yellow-pine forests, of Nevada, Utah, Arizona, New Mexico and northern Mexico are the habitat of the Cliff Chipmunk.

5 *QUADRIVITTATUS* GROUP

COLORADO CHIPMUNK (*Eutamias quadrivittatus*)

Yellowish tinges the gray color of this chipmunk's head, sides and rump. Its tail is yellowish underneath and fringed with yellow or whitish hairs.

Colorado Chipmunks are commonly found in forests and brushy areas in mountains of Utah, Colorado and northern Arizona and New Mexico.

RED-TAILED CHIPMUNK (*Eutamias ruficaudus*)

Large and brightly colored, the Red-tailed Chipmunk's shoulders and sides are yellow, its rump is gray. Its light dorsal stripes and ear patches are grayish white; its tail is dark orange-yellow.

Red-tailed Chipmunks are found in southern British Columbia, northeastern Washington, northern Idaho and northwestern Montana.

GRAY-COLLARED CHIPMUNK (*Eutamias cinereicollis*)

Pale gray fur on sides of neck and shoulders distinguishes this chipmunk. Its dark gray pelage has a yellowish tinge on the sides of its body. A median black stripe and dark brown lateral stripes run down its back.

Gray-collared Chipmunks occupy a relatively small range that extends from central Arizona into New Mexico and the Guadelupe Mountains region of Texas.

LONG-EARED CHIPMUNK (*Eutamias quadrimaculatus*)

The Long-eared Chipmunk, large and dark grayish or tawny in color, has contrasting body stripes, conspicuous white patches behind its relatively long, slender ears and a tail that is reddish brown beneath and fringed with white-tipped hairs.

Its small range is in northern and central Sierra Nevada in California and in western Nevada in the vicinity of Lake Tahoe. Open places in yellow-pine and Douglas fir forests, where fallen logs offer shelter among manzanita bushes or other brush, are typical habitat of the Long-eared Chipmunk.

LODGEPOLE CHIPMUNK (*Eutamias speciosus*)

Slightly larger size (total length seven and three-quarters to eight and three-quarters inches), more contrasting light and dark stripes distinguish the Lodgepole Chipmunk from the Yellow-pine Chipmunk. Its outer light stripes are white (rather than yellowish tinged) and broader than its inner light stripes. Sometimes its outer dark stripes are nearly obsolete.

Coniferous forests, especially lodgepole pine, of the central and northern Sierra Nevada in California and extreme western Nevada are the home of Lodgepole Chipmunks. Where other chipmunk species overlap its range, the species occurs only in or near dense stands of lodgepole pine. Lodgepole Chipmunks range south in the San Bernardino and San Gabriel ranges and occur on Mount Pinos in Ventura County, California.

PANAMINT CHIPMUNK (*Eutamias panamintinus*)

This small brightly colored chipmunk has a gray head and rump and reddish fur on its shoulders and back. The median stripe is dark brown, the other dark stripes are reddish or grayish, while the stripes flanking the outer white stripes are faint or nearly obsolete.

Panamint Chipmunks live among the boulders and on cliffs in piñon-juniper areas in eastern California and western Nevada.

UINTA CHIPMUNK (*Eutamias umbrinus*)

This chipmunk closely resembles the Colorado Chipmunk. Top of head and neck are grayish; outer light stripes are pure white; outer dark stripes are faint. The Uinta Chipmunk also resembles the Lodgepole Chipmunk; study of the bacula of species of *Eutamias* shows the Uinta Chipmunk to be a distinct species.

Uinta Chipmunks live in coniferous forests from 7000 to 11,000 feet elevation in northwestern Wyoming, Nevada, Utah, Colorado and northern Arizona and New Mexico.

PALMER'S CHIPMUNK (*Eutamias palmeri*)

In its winter pelage this species has indistinct dorsal stripes. Palmer's Chipmunks differ from nearby populations of Uinta Chipmunks, in having more reddish colored dark dorsal stripes and more tawny color on the underside of the tail.

Palmer's Chipmunk occurs in the area of Charleston Peak, Clark County, Nevada. Hall writes:

> Isolated on Charleston Peak, this chipmunk has developed characters which differentiate it . . . from its near relatives of the species *E. umbrinus* . . . It ranges from 7000 feet altitude in the yellow pine belt up to timber line at about 12,000 feet.

BULLER'S CHIPMUNK (*Eutamias bulleri*)

Rather broad, blackish facial markings, a gray collar, dark brownish body stripes, brownish sides and reddish underside of the tail characterize Buller's Chipmunk. The species is found in northern Mexico.

II

THE TREE SQUIRRELS

Long bushy tails, absence of dorsal stripes, spots or flecks, and lack of internal cheek pouches distinguish the tree squirrels from their family relatives, the ground squirrels and chipmunks. *Sciurus,* the Latin word for "squirrel," is derived from the Greek words *skia* meaning "shadow" and *oura* meaning "tail" and is translated as "shade tail" or "creature that sits in the shadow of its tail." This the gray squirrels quite literally do. *Tamias* is a Greek word meaning "one who lays up stores," referring to the hoarding habits of Red Squirrel and Douglas Squirrel *(Tamiasciurus).*

Tree squirrels' skulls exhibit greater width between the eye sockets and greater depth of braincase than do skulls of ground squirrels or chipmunks. Although cranial characters are markedly similar among the tree squirrels, skull features and differences in dentition cause taxonomists to place North American tree squirrels of the genus *Sciurus* in five subgenera.

The subgenus *Neosciurus* includes the Eastern Gray Squirrel and some eleven species of Mexican and Central American squirrels. Of this assemblage Hall and Kelson write:

> . . . the taxonomic status and phylogenetic affinities are poorly known . . . the 12 nominal species probably belong to 6 or possibly 7 species at most.

The Western Gray Squirrel, placed by itself in the subgenus *Hesperosciurus,* has a broad skull and relatively "massive" molar teeth. *Otosciurus* is the subgenus that contains the elegant, tassel-eared Abert's and Kaibab squirrels. The Fox Squirrel, largest of the North American tree squirrels, the Arizona Gray Squirrel, the Chiricahua Squirrel and four Mexican relatives are grouped in the subgenus *Parasciurus,* while two Central American species form the subgenus *Guerlinguetus.*

Two genera of small tree squirrels occur in Central America: the montane squirrels (*Syntheosciurus*) and the dwarf squirrels (*Microsciurus*).

Scolding and sputtering characterize *Tamiasciurus* (Red Squirrel and Douglas Squirrel). Sentinels of the forests, they make up for their small size by being the noisiest of the squirrels.

All these arboreal acrobats are highly specialized for tree living. They have plume-tails for balance, powerfully muscled hindquarters for leaping and sharp claws for clinging. These squirrels depend upon trees for shelter, for food and for the network of routes their branches provide.

GRAY SQUIRRELS AND THEIR RELATIVES

EASTERN GRAY SQUIRREL (*Sciurus carolinensis*)

Description. A bushy tail with long, slightly crinkled hairs that are banded blackish and tan and broadly tipped with white, measures eight or nine inches of the sixteen to twenty-three inch total length of this squirrel. Its salt-and-pepper gray coat is made up of gray underfur; guard hairs are gray basally, banded with buff-brown, then black and tipped with white. White-tipped guard hairs on the tail distinguish the Eastern Gray Squirrel from the Fox Squirrel, which has rusty red guard hairs on its tail. On each side of the upper jaw are two premolars, the first peg-like; the Fox Squirrel has a single premolar. Gray squirrels lack the conspicuous ear tufts of their relatives, the Abert's and Kaibab squirrels.

In winter the soft, light, gray fur grows long and dense. The winter pelage becomes worn in spring and the yellowish brown fur that covers parts of the squirrel's head, shoulders, back and upper surfaces of the feet is more noticeable. Underparts are whitish. The ears are backed with white fur, the eyes ringed with white or buff.

Albino and melanistic gray squirrels occur—sometimes these mutations, either partial or complete, are so common that they become the dominant color phase in an area. Black individuals, varying from brownish to jet black, in some localities outnumber the grays. According to Leonard Lee Rue III:

> Occasional black squirrels may be found throughout the gray's entire range. The same is true of the albinos, except that white squirrels frequently become common or dominant only in isolated pockets. Trenton, New Jersey, has quite a few albino squirrels. Greenwood, South Carolina, has a colony that inhabits about 100 acres. The largest concentration of albinos, over 1,000, can be found in Olney, Illinois. This particular strain can be traced back to 1892, coming from a pair that was owned by a saloonkeeper.

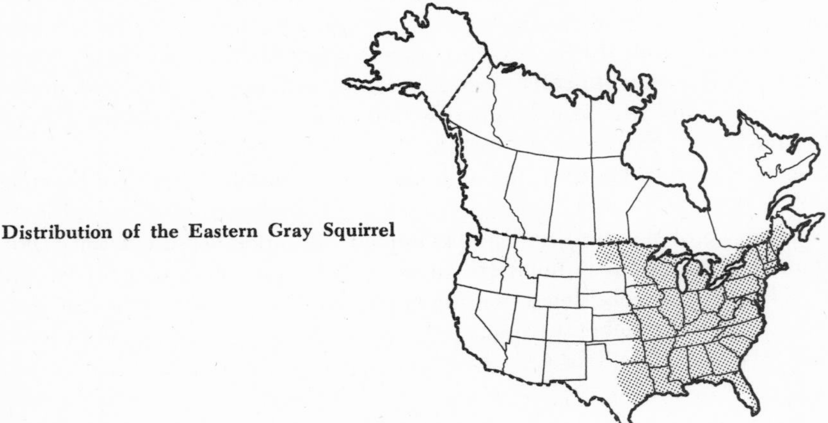

Distribution of the Eastern Gray Squirrel

Distribution. From southern Canada to the Gulf of Mexico the range of the Eastern Gray Squirrel blankets most of the eastern half of the United States, from eastern Saskatchewan and eastern Texas to the Atlantic Coast.

Dense forests of mature hardwoods originally covered the entire eastern part of the United States, as far west as the high grass prairies. Seton estimates that within this one-million-square-mile range more than a billion gray squirrels once lived. Hardwood forests that remain are still the home of the Eastern Gray Squirrels. They also live in city parks and suburbs, where large nut and shade trees offer food and denning sites. Oaks, beeches and hickories are preferred, but frequently gray squirrels live where these trees are mixed with other hardwoods and, sometimes, pines and other conifers.

Like the Fox Squirrel, the Eastern Gray Squirrel has been introduced into the Pacific states and can be found in numerous city parks. Gray squirrels have also been introduced into Britain and South Africa.

THE WAYS OF AN EASTERN GRAY SQUIRREL

The gray squirrel awoke at dawn and peered out of her nest. Several days after the man who lived in the house had climbed up a ladder to fasten the flicker box onto the trunk of the huge white oak, the gray squirrel had moved in. She gnawed the box's round entrance hole to make a more comfortable doorway. Oak leaves, turned brown by autumn frosts, were gathered, mouthfuls at a time, from the tree's branches and carried into the flicker box for nesting material. Den trees were scarce in the tree-shaded suburb where the gray squirrel lived. The flicker box was a welcome substitute for a den high up in a big tree. Most of the gray squirrels in the area lived in leaf nests the year around. These were built close to the trunk of a tree, in a fork or on a strong limb, and were located thirty to fifty feet above ground. Most of the nest-building and nest-refurbishing activity took place in autumn. Twigs with leaves attached were used to form a supporting platform, then more leaves were added and roughly woven with dead twigs. On this was a compacted base of decaying organic matter. The outer shell of leaves and twigs covered an inner layer of closely woven material, which surrounded a nest cavity twelve to sixteen inches in diameter. Some of the squirrels appeared to be less industrious and simply heaped bark, moss and grass onto platforms of twigs, then hollowed out nests for themselves. Insects almost always inhabit squirrel leaf nests. Nests built in the fall deteriorate the following summer when leaf-eating larvae skeletonize the leaves.

Out on a large horizontal limb of the oak the gray squirrel groomed herself in the dim early light. She paid special attention to her tail and rubbed her forepaws over her face. Dropping to all fours, she shook herself, stamped, shuffled with her feet and yawned. Then she started down the oak's trunk head first, her body flat against the bark, her legs spread out sideways, her head held a little up and out. Her fluffed out tail was held straight behind. Sharp, hook-like claws, an important adaptation to the squirrel's aboreal way of life, made her descent jerky until a few feet from the ground. With a bound the squirrel landed on the grassy hillside and began, in zig-zag hops, to look for food. Like most gray squirrels she would spend most of her life in and around her nest tree, familiar with travel lanes through the treetops, escape routes and food sources. A study of gray squirrels in two Maryland woodlots revealed that the animals occupied relatively small overlapping home ranges, 0.2 to 7.2 acres in extent. Some squirrels shift their homes, as different foods attract them, and move over as much as a five-mile area. But the gray squirrel had an abundant food supply where she lived. When winter stores were depleted there was always the bird feeder to be raided.

Acorns, hickory nuts, walnuts and butternuts supply much of the gray squirrels'

EASTERN GRAY SQUIRREL
(*Sciurus carolinensis*)

diet from late summer to the following spring. Individual squirrels consume about two pounds of food a week, or 100 pounds a year. In seasonal sequence the squirrels feed on wild fruits, nuts and berries. Bark and twigs often keep squirrels alive when food stores run out in late winter. Early spring produces sweet sap to lick, swelling buds of maple and elm and tender opening twigs of oak. The squirrel bites off a terminal twig, holds it and nips off and nibbles a few buds. Then, dropping that twig, it reaches for another. John T. Nichols writes:

> If every nut that was planted were presently dug up and consumed, if seeds were not carried about in times of plenty and then tossed aside, these squirrels would be less helpful agents in spreading the forests on which they depend.

Summer adds fruits such as wild grape, pokeberries, hackberries, wild cherries, berries of dogwood and viburnum and skunk cabbage seeds to the squirrels' fare. Mushrooms, insects and their larvae and occasionally eggs or a nestling bird are eaten, or a bit of carrion sampled. Like other rodents, gray squirrels are fond of gnawing on shed antlers or bones. One gray squirrel was observed to keep a favorite, well-gnawed bone in a small tree cavity.

The acorn crop of the big white oak had ripened in late August—autumn became a time of plenty. Squirrels concentrated at the best feeding places; burying activity was at its height. The gray squirrel gathered acorns by perilously balancing among the tree's thin branches. Often she would hang by her hind feet to secure acorns. Then early frosts scattered the acorns. The gray squirrel in her zig-zag morning ramble picked up an acorn in her mouth, sat back on her haunches and held the bit of breakfast in her forepaws and opened it with her strong, narrow, sharp-pointed incisors. After consuming several acorns the squirrel began to select acorns to store—it seemed she would never put away enough for the winter. Cutting the cup from the acorn, she carried it off fifty to one hundred feet from the oak. With her forepaws she quickly dug a hole an inch and a half deep, put the acorn in it and pushed soil, leaves and twigs over it. One to three minutes was spent on each acorn thus cached. Gray squirrels have been known to cache food in hollow tree trunks or fallen logs. Squirrels in the University of Wisconsin Arboretum cache butternuts in natural pockets formed by close branches of the uppermost whorls of pine saplings.

A suspicious noise interrupted the gray squirrel's foraging. She stopped and sat bolt upright, her front paws pressed against her chest. Her large eyes peered intently. Then she detected the stealthy movement of a cat, crouching beneath a clump of forsythia. Heavy muscles of hind legs and back carried her in bounds and a final five-foot leap to the large oak. She scampered to the far side of the trunk and scaled upward. Reaching the security of the horizontal limb near her nest she paused, flicked her tail rapidly, scolded in a high-pitched voice and chattered with her teeth. The cat soon left its hiding place in the forsythia.

Of all the gray squirrel's adaptations her plume-tail, often flicked and fluffed out, was the most conspicuous. Used as a balance in jumping, climbing, running along branches and making quick turns, it also acts as a parachute should the squirrel fall. Flicked from side to side, the tail can be used to distract a pursuing predator. Sometimes a squirrel escapes a predator's grasp by losing the skin and hair tail sheath or even some of its tail vertebrae. The exposed vertebrae dry up and fall off. Squirrels with partial tails are not uncommon. White hairs usually grow in to accentuate the loss of a tail tip. The tail also serves as a rudder if a squirrel is forced into water. On hot, sunny days the tail makes a useful shade, and in cold weather it can be wrapped around the squirrel's body for warmth. Of the squirrels' habit of flicking and fluffing their tails Monica Shorten writes:

> The muscles of the tail can be used to alter the position in which the tail hairs lie; when the tail is spread out, the muscles pull the hairs so that they lie at right angles to the tail-bone, instead of lying backwards on each side, and a "fire-brush" effect is given when the muscles pull the hairs up and down as well as forwards.

The gray squirrel stretched out on the horizontal limb. Like others of her kind she was most active during the several hours following sunrise. Most of the day that remained she would spend in her nest, or lying on the limb, sunning and dozing. At mid-day and for an hour or so before dusk she would again be active.

By late September all the young of the year have been weaned. Squirrel numbers are at their highest annual level, and it is time for autumn dispersal. Shorten writes of this time of year:

> There is still an abundance of food, unless some important mast crop has failed; and dry clear days with even temperatures and little wind encourage wide ranging and longer periods of activity. The stage is set for the annual re-shuffle. . . . Motorists may notice that more squirrels are being killed on the roads . . . and small woods are full of squirrels one day and deserted the next.

Dispersal occurs regardless of the size of squirrel populations and the amount of food available, but in some years there are large-scale emigrations. The female and her gray squirrel neighbors were busy caching an abundant acorn crop that fall, but squirrels in some rural parts of the northeast were affected by an acorn crop failure. Unable to find enough acorns, the squirrels quickly devoured the other nut crops. Then they began to forage farther and farther afield. E. M. Reilly, Jr., writes:

> Generally . . . the mature well-established squirrels retain their territories and drive out unestablished youngsters who become nomads searching aimlessly for food and shelter.

If the nut crop failure is widespread, the hungry squirrels cannot find food in nearby woodlots or over the next hill. In some years oaks fail to bear enough acorns, and hickories, walnuts, beeches and hazels also produce a meager nut crop, perhaps because drought, cold or too much rain interfered with development of the fruit. In desperation the squirrels attempt to cross busy turnpikes and highways—some make it; the bodies of others litter the roadsides, mute testimony of an autumn emigration of squirrels, a time of permanent departure of large numbers of adults and young from the home areas. H. H. T. Jackson quotes a Mr. Broach, who witnessed a "continuous movement" of gray squirrels near Pepin, Wisconsin, in the fall of 1914 or 1915. The squirrels swam the quarter-mile span of the Mississippi River to the Minnesota shore at the average rate "of two entering and two emerging from the water every few minutes for about a day." W. J. Hamilton Jr., gives this account:

> A squirrel emigration of no mean proportion occurred in Connecticut and New York during the fall of 1933. More than a thousand squirrels were observed swimming the Connecticut River between Hartford and Essex, a distance of about forty miles. Many of them became exhausted and drowned, while others climbed up on logs or drift on which they rested as they floated down the river. Still others came aboard ferry boats or scampered across bridges which span the Hudson River, all moving in a westerly direction. Again, during the fall of 1935, a similar and more extensive movement of these rodents was noted, the emigration extending into western New York.

More recently A. W. Schorger reports a 1946 emigration of both gray and Fox squirrels in northwestern Wisconsin. Squirrel populations were estimated at Sarona, Wisconsin to number fifty per square mile. Approximately half the squirrels appeared to be on the move. From mid-August on, squirrels of both species were observed swimming in lakes and rivers, crossing highways and invading prairie cornfields. That year a very poor acorn crop followed a year when acorns were abundant and squirrel populations were relatively high. According to William H. Longley, Minnesota gray squirrels were on the move in 1958, following failure of the acorn crop and at a time when there was a high squirrel population. September of 1960 saw a general north-to-south emigration of gray squirrels in southeastern Massachusetts and eastern Connecticut. In the fall of 1968 squirrel bodies littered both banks of the Hudson River. Reilly writes:

> Driven by the hope that there are more nuts "on the other side," squirrels battle the currents and may pass an equally forlorn squirrel swimming in the opposite direction.

In spite of the preceding fall's squirrel population density, the carrying capacity of an area remains fairly constant year after year. All the reports of squirrel emigrations indicate that high proportions of the squirrels found dead are young animals. Richard G. Van Gelder notes the biological significance of such emigrations. Dispersal by the young squirrels reduces the population in the parental environment, thus contributing to a more or less stable home population. If the species population is at a low ebb the dispersing animals will repopulate the area, or if the population is high the dispersing squirrels are forced out into new areas where they may become established, thus increasing the range of the species. Van Gelder also points out:

> The great mortality that occurs during dispersal . . . reduces the pressure of predators on the parents, and it introduces a strong natural-selective pressure on the young dispersing animals.

There are numerous early accounts of mass excursions of gray squirrels, movements involving tens and even hundreds of thousands. Seton estimates that nearly half a billion squirrels took part in the 1842 emigration that occurred in southeastern Wisconsin. And there are records of enormous squirrel "kills," for the abundant animals frequently destroyed the crops the early settlers struggled to plant in woodland clearings. Squirrel numbers declined as more and more forest land was cleared by logging and burning. During the 1800's relentless war was waged against gray squirrels. Many states offered bounties, and Ohio declared that county taxes might be paid in squirrel "scalps," valued at three cents apiece. Cahalane writes:

> From about 1900 to 1915, conservationists feared that the eastern species of gray squirrel might become extinct. However, reasonable protection of wildlife and forests has permitted it to reoccupy suitable habitats in some numbers.

Today many a woodsman can determine how good the ensuing squirrel hunting season will be by assessing the acorn crop. Living as she did in the great oak that spread over the slope below the small house, the gray squirrel did not fear people. In fact she almost trusted them, although she was wary of the dog and two cats that shared the peoples' house. But woodland gray squirrels are far more wary and easily frightened. The alarm call of one squirrel, a rapid *kuk, kuk, kuk,* warns others of an intruder. Sometimes several squirrels join in a scolding chorus to taunt a predator. The approach of man, marked by snapping twigs and rustling leaves, sends them fleeing through the treetops. Or a frightened squirrel may flatten out against the upper trunk or along a branch and "freeze." Merriam gives this account of the wary gray squirrel:

> Let someone try to get within gunshot and observe the result. His very approach seems to render them invisible. Those that were near their holes have disappeared within, and the others are hiding behind the trees upon which they were sporting when the enemy appeared. As he advances they rotate slowly about the trunk, always keeping on the farther sides so that the body of the tree remains between them. Even if he knows that a squirrel is on a certain tree, it is doubtful if he gets a shot. A momentary glimpse of its ears or a part of its tail constitutes all he is likely to discover as he walks round the tree.

As to the preparation and cooking of gray squirrels, Shorten gives six recipes, and adds:

> Young squirrels are the pick of the bag, . . . Old squirrels, especially during the breeding season, have a characteristic taste and smell which can be neutralized by soaking the joints overnight in water with vinegar or salt, or filling the body cavity with sliced apple which is removed before cooking.

During the fall the female gray squirrel shed her summer coat for thick, silver-gray winter pelage. She grew conspicuously fat. A good layer of fat would enable her to remain in the nest for longer periods in cold winter weather. Gray squirrels that lived not far away, on the city green, never lost their scraggly appearance. They depended on handouts of bread, peanuts and popcorn.

Snow and cold weather came. The gray squirrel left her nest in the flicker box only to forage for her caches, or to gather additional nest material from the dead brown leaves that clung to the oak's branches. One of the winter's storms developed into a blizzard, keeping the drowsy squirrel in her nest for almost three days. Some gray squirrels' nests sheltered two or three squirrels, huddled together for warmth. Because they had fed well that autumn and acquired good layers of fat, the squirrels' caches were more than adequate that winter. The animals shared each others' buried stores. Small, neatly dug holes in the snow marked their unearthed caches. Memory probably took the squirrels to the caching areas, but it was their keen sense of smell that located each buried nut. The bird bath near the rhododendrons, the gray squirrel's water source for much of the year, was frozen solid. But the squirrel quenched her thirst by eating small amounts of snow and in spring by licking maple sap.

One especially cold morning the gray squirrel was joined at the feeder by a shivering squirrel. Fur was gone from his neck and shoulders. He suffered from scabies, or mange, caused by the *Sarcoptes* mite, a condition that often plagues squirrel populations in winter and early spring. The mites burrow into the skin, scabs form and hairless patches appear on the body. Because of the loss of fur and generally weakened condition that prevents active foraging, scabies-afflicted squirrels often become easy prey or succumb to freezing temperatures; some squirrels recover and grow new hair. Gray squirrels also suffer from a host of other parasites: ticks, lice, fleas, flies and botfly larvae, roundworms, tapeworms and protozoa.

December brought unrest among the male gray squirrels. Their shrill buzzing calls pierced the air. Frequently they battled among themselves or chased after female squirrels. More often than not the female would turn on a pursuing male with sharp cries and vicious bites and drive him from the limits of her nest tree. Shorten gives this description of gray squirrel mating chases and the courtship that sometimes ensues:

> . . . groups of excited males pursue a female which displays real or simulated reluctance. Sometimes one or more of the group will lose footing and fall to the ground during the scramble through the trees. When a male is rebuffed he often relieves his pent-up emotions by gnawing vigorously at a branch or a patch of bark; . . . During courtship [the male] prowls restlessly before [the female], his tail fluffed out stiffly, jerking and vibrating with nervous excitement; he chatters his teeth, and strikes his forefeet down on the branch . . .

For a few days in January such displays occurred between a handsome big male gray squirrel and the female who lived in the big oak, then he was gone.

The gestation period of gray squirrels is forty-four to forty-five days. So it was in early March that the female gave birth to four half-ounce babies, pink, naked and with their eyes and ears sealed shut. The female was fiercely protective about her nest now. The baby squirrels doubled their weights in the first week, and for the first two weeks they lay curled asleep in the dark warmth of the nest, awakening only to feed. In the third week short, dense fur began to cover their bodies; gradually their ears and eyes opened.

One of the baby squirrels tumbled from the nest. The mother squirrel would have been quick to rescue him, but she was out foraging. His cries attracted a cat, and the baby squirrel was carried off. The cat belonged to a little girl who retrieved the unhurt squirrel from the cat's jaws. Not knowing where the baby squirrel's nest was, the little girl became its foster mother. The squirrel fed eagerly from a small puppy nurser and spent much of its time sleeping in a pocket of the child's dress. For rearing of orphan squirrels Lee S. Crandall gives this advice:

> Young eastern gray squirrels are easily reared on whole cow's milk given with a medicine dropper or very small rubber nipple. Vitamin concentrate drops and precooked baby food, such as Pablum, can be added. When ready to feed themselves, the youngsters can be started on bread and milk followed by soft fruits and vegetables, and soon can be brought gradually onto the adult diet. . . . It is presumable that young squirrels of most species will give similar results.

Soon the three remaining youngsters were ready to leave the nest for the first time. One warm spring day the big oak tree was alive with squirrels—the seven-weeks-old squirrels, well-furred and nearly nine inches long, had come out to explore its branches. Although the little squirrels still suckled, they sampled some of the foods their mother ate. By ten or eleven weeks of age they would be fully weaned. Gradually the young squirrels adjusted to life outside the nest. During the period of dependency the squirrel family foraged together. Often one of the young squirrels would pause, run to its mother or a litter-mate and sniff and nuzzle its flanks. Possibly the nuzzling represents the young squirrels' transition from olfactory recognition, used in the nest, to visual and auditory recognition, so important in adult squirrel life. J. P. Hailman writes of the behavior of young squirrels:

> Just as the young squirrel must learn what is "climbable," so it must learn what is "edible," and foraging during the dependency period may provide that experience. Besides the selection of food, other feeding behavior, such as the opening of nuts, improves during this time.

Some of the female gray squirrels bred again that summer, producing August litters that would remain through the winter with their mothers. Leaving their first-litter youngsters and their nests, these females moved into other nests, usually nearby, to give birth to their second broods. But the female gray squirrel that lived in the flicker box produced only one litter that year.

A nest hole in a tall oak not far away was home to a raccoon and her young. Since there were no foxes, bobcats, weasels or snakes in the area, the raccoons, neighborhood cats and dogs and, rarely, a hawk were the only predators that concerned the gray squirrel. Even a tree-climbing 'coon would find it difficult to get at her family when they burrowed into the bottom of their nest in the deep flicker box. In rural areas black rat snakes and other tree-climbing snakes are deadly enemies of gray squirrels, along with others such as red-tailed, red-shouldered, broad-winged, marsh and Cooper's hawks, goshawks, great-horned, barred and long-eared owls.

The young gray squirrels moved out of their flicker box home in early fall.

They would not be fully grown until their second year, although they might breed as yearlings. Six to ten years would be their potential life span, an allotment reached by few squirrels. Captive gray squirrels have lived as long as fifteen years.

GRAY SQUIRRELS IN MEXICO AND CENTRAL AMERICA

Gray squirrels in Mexico and Central America are distinguished from other Mexican and Central American tree squirrels by their medium size, predominately gray backs, the presence of the tiny peg premolar on either side of the upper jaw and absence of ear tufts.

A. Starker Leopold writes of their habitats:

> Most of the gray squirrels in Mexico live in tropical forests that have been fairly recently opened up by clearing. The combination of remnants of native forest interspersed with cornfields is actually very favorable habitat for gray squirrels—better, perhaps, than the virgin forest—and may account for the abundance of these animals in some coastal areas. The upland species of gray squirrels (*poliopus, nelsoni* and *griseoflavus*), however, which live in the central and southern pine-oak lands, have had their homes destroyed by clearing and forest depreciation, and these species are relatively scarce.

Habitat destruction has in turn affected the systematic relationships. Hall and Kelson have this to say:

> Much of the formerly forested parts of Mexico where these squirrels occurred is now denuded of trees, and the natural geographic distribution of the animals is much altered by constriction and fragmentation. Intergrades between certain morphologically distinct populations probably no longer exist and it is likely that the systematic relationships of the mid-American squirrels can never be satisfactorily resolved except inferentially.

RED-BELLIED SQUIRREL (*Sciurus aureogaster*)

Description. Grizzled gray upperparts marked by patches of reddish or brownish fur on neck and rump, or only on shoulders, distinguish the Red-bellied Squirrel. Its belly is reddish; its tail blackish grizzled with white.

Distribution. Humid, lowland jungles to open pine forests at high altitudes are home to this highly arboreal squirrel. Feeding on such fruits as mangos, green figs, jubo plums and tamarind pods, it seldom is seen on the ground. Its range in central Mexico borders the Gulf of Mexico and includes the states of Tamaulipas, San Luis Potosi, Hidalgo, Veracruz, Puebla, Oaxaca, Tabasco and Chiapas, Mexico and northwestern Guatemala.

RED-BELLIED SQUIRREL
(*Sciurus aureogaster*)

MEXICAN GRAY SQUIRREL (*Sciurus poliopus*)

Description. Tawny or brown neck and rump patches usually decorate the yellow- or rusty-tinged grizzled gray fur of this squirrel. The underside of its gray tail has a median rusty or tawny stripe.

Distribution. Mexican Gray Squirrels are found in pine forests and in deciduous woodlands in parts of the states of Nayarit, Jalisco, Colima, Michoacan, Mexico, Guerrero, Puebla and Oaxaca. According to Leopold, the Mexican Gray Squirrel, the Red-bellied Squirrel and Deppe's Squirrel are known to make altitudinal migrations in search of better feeding grounds. Their vertical ranges extend from 4,000 to 6,000 feet, where they feed on wild figs and acorns, to as high as 12,000 feet, where pine cones are gnawed for their seeds. These same squirrel species are sometimes guilty of crop damage.

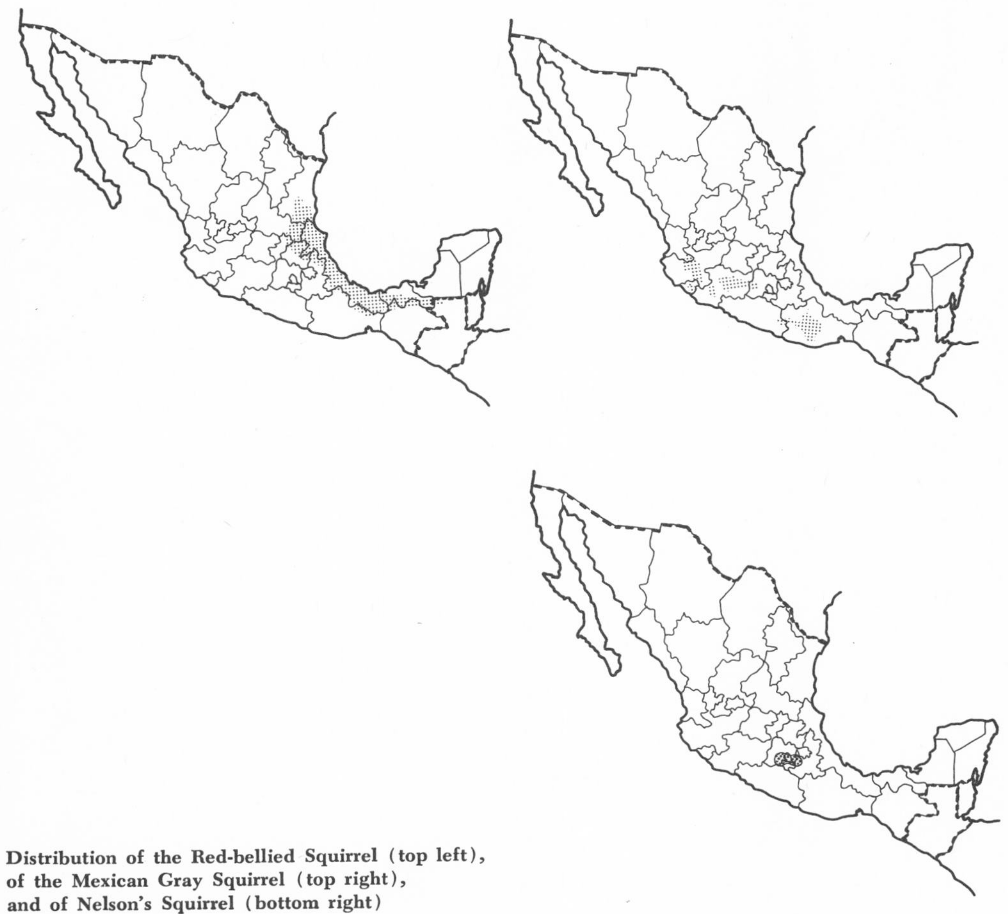

Distribution of the Red-bellied Squirrel (top left), of the Mexican Gray Squirrel (top right), and of Nelson's Squirrel (bottom right)

NELSON'S SQUIRREL (*Sciurus nelsoni*)

Description. Probably because populations of Nelson's Squirrels live in isolation of high altitude forests their color is highly variable. Their dark coat color may be sooty gray or blackish brown, tinged with yellow or buff. The belly is blackish. A ventral median tail stripe may be tawny or rusty.

Distribution. Nelson's Squirrel occurs in the mountains of the Mexican states of Mexico, Puebla, Distrito Federal and Morelos.

COLLIE'S SQUIRREL (*Sciurus colliaei*)

Description. Yellowish grizzled gray above, white below, this squirrel has a tail that is blackish above and grayish or blackish yellow below and bordered with white.

Distribution. Lowlands of the west coast of Mexico, including the states of Nayarit, Jalisco and Colima.

SINALOAN SQUIRREL (*Sciurus sinaloensis*)

Description. Reddish yellow fur mixed with black covers the upperparts of this squirrel; its underparts are whitish.

Distribution. Continuous with the range of Collie's Squirrel, the range of the Sinaloan Squirrel includes lowlands of Sinaloa, Mexico.

SONORAN SQUIRREL (*Sciurus truei*)

Description. Dark yellowish grizzled with black fur distinguishes the Sonoran Squirrel. Its fur is white below and its tail is grizzled black and buff with a white wash.

Distribution. Fields and woods of Sonora, Mexico, are the habitat of this squirrel.

SOCIABLE SQUIRREL (*Sciurus socialis*)

Description. The fur of the Sociable Squirrel is whitish gray with a yellowish tinge. Often there is a neck patch and a rump patch of yellow or rufous fur. Its belly is reddish. Its name derives from its social habits.

Distribution. Lowland scrub forests, and sometimes cornfields or cocoanut groves, of Guerrero, Oaxaca and Chiapas, Mexico.

GUATEMALAN GRAY SQUIRREL (*Sciurus griseoflavus*)

Description. Hall and Kelson write of this squirrel:

> In "average" pelage S. *griseoflavus* is distinguishable by heavy body, and the combination of grizzled yellowish-brown back with reddish brown venter, and

Left: Distribution of the Collie's Squirrel (A), Sinaloan Squirrel (B), Sonoran Squirrel (C). Right: Distribution of the Sociable Squirrel (A), Guatemalan Gray Squirrel (B), Yucatan Gray Squirrel (C)

lack of patches on nape and rump. . . . Because of its large size, delectable meat, and extreme wariness, this squirrel is a highly esteemed game animal.

Distribution. Pine and oak forests of Chiapas, Mexico and Guatemala.

YUCATAN SQUIRREL (*Sciurus yucatanensis*)

Description. The thin coarse fur of this squirrel is grizzled black and gray with a yellow tinge. Sometimes it has "dingy white" ear tufts.

Distribution. Yucatan Squirrels are found throughout the Yucatan Peninsula, in Campeche, Yucatan, Quintana Roo, Mexico and in parts of Guatemala and British Honduras.

VARIEGATED SQUIRREL (*Sciurus variegatoides*)

Description. Faint rings sometimes appear on the blackish tail of the Variegated Squirrel. Its coarse, bristly fur varies in color from blackish to grizzled yellowish gray above and from white to cinnamon buff below. If its underparts are not white there are usually patches of white fur.

Distribution. This squirrel is abundant in humid lowlands and in dry woodland situations from Chiapas, Mexico and Guatemala south throughout El Salvador, Honduras and Nicaragua, and occurs also in Costa Rica and Panama.

DEPPE'S SQUIRREL (*Sciurus deppei*)

Description. This small, dark squirrel has upperparts that are grizzled rusty brown, olive brown or grayish brown. The tail is black washed with white above and yellow or rusty below, and the underparts are white, yellowish or rufous, often with patches of buff fur.

Distribution. Deppe's Squirrel is at home in the dense vegetation of humid lowland rain forests and tropical evergreen forests. At higher elevations in Oaxaca and Chiapas it occurs in areas of cloud forest. Its range includes Veracruz, parts of San Luis Potosi, Hidalgo, Puebla and Oaxaca, Tabasco, Chiapas, the Yucatan Peninsula in Mexico, and Guatemala, central Honduras and Nicaragua. Leopold relates this description of Deppe's Squirrels in a Chiapas cloud forest:

> Among the tall pine trees, heavily grown with lianas and epiphytic plants, it was so hard to see the animals that collecting specimens was difficult. . . . The active little scamps would dash to safety in the dense tangles of the tallest trees, there to crouch quietly until the danger had passed.

Deppe's Squirrels have been reported to congregate at times, "swarming through the trees like locusts." They gnaw the pods of ebony trees for their seeds and eat the fruits of such trees as the chicle zapote and the amate.

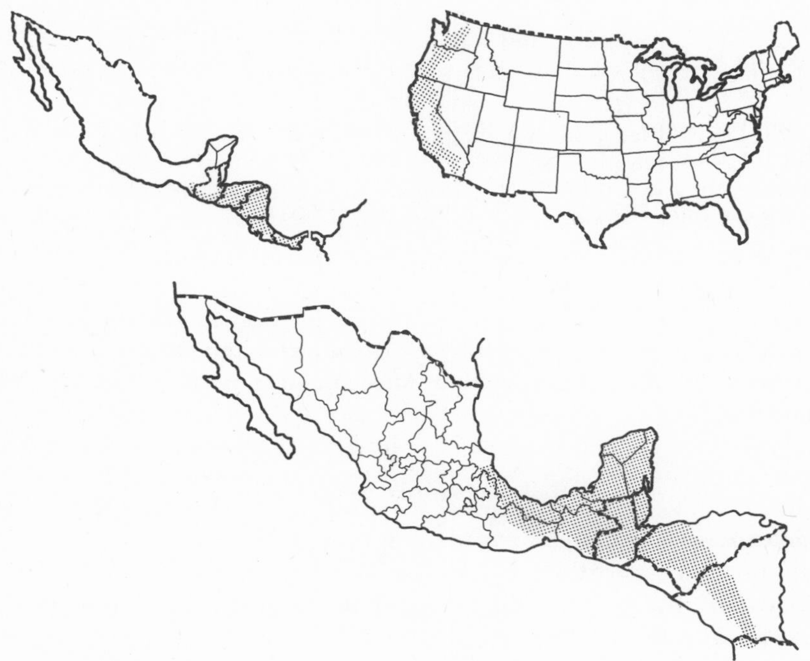

Distribution of the Variegated Squirrel (top left), Western Gray Squirrel (top right), and the Deppe's Squirrel (bottom)

WESTERN GRAY SQUIRREL (*Sciurus griseus*)

Description. Larger size (total length is twenty-two to twenty-four inches), uniformly blue gray upperparts, white underparts, dusky feet and a conspicuously broader tail that is edged with white distinguish the Western Gray Squirrel from its smaller and browner relative the Eastern Gray Squirrel.

Distribution. From central and western Washington south through western Oregon and in California south in the Coast Ranges and along the Sierra Nevada to northern Baja California. Western Gray Squirrels are at home in redwood forests as well as in mixed stands of oak and pine, or ponderosa pine and fir, between 3,000 and 8,000 feet. Foothill areas are visited in the fall when acorns ripen. Except for family groups it is usual to see only two or three Western Gray Squirrels at a time.

Habits. Like their eastern cousins, Western Gray Squirrels are arboreal acrobats, capable of bounding along small branches and leaping through space, using their plume-like tails to maintain balance. Surpassed as aerialists by Douglas Squirrels, Western Gray Squirrels do sometimes jump from tree to tree without hesitation and in forests where trees are close together they often follow arboreal routes. According to Bachman, the Western Gray Squirrel prefers ground travel. He writes:

> It is exceedingly swift on the ground, and will not readily take to a tree; or if it does, ascends only a few feet, and then jumping down to the ground, runs off with its tail held up, but curved downwards towards the tip like that of a Fox when in flight.

Much time is spent foraging on the ground, where the squirrels must be ever-watchful for their enemies: coyote, fox, bobcat, hawks and owls.

Because they are primarily acorn-eaters, Western Gray Squirrels are most often found in open forests containing any of various kinds of oaks. According to Ingles, the Western Gray Squirrel:

> Sometimes . . . extends its range up the mountains . . . where there are no oaks and where the deadly enemy of all tree squirrels, the marten . . . lives. Here the winters are long, the snow may reach 15 feet on the level and stay on the ground for months. In such an environment the western gray squirrel may last for a year or two, and even raise a litter of young. When the snow becomes deeper, the marten comes lower down the mountains to find food. It can easily catch this slow-moving squirrel in the trees or in the enlarged hole it uses for a winter nest.

Bulky nests, two feet in diameter and three feet high, built of sticks and shredded bark, are conspicuous in the treetops where Western Gray Squirrels occur. Sometimes an abandoned crow's or hawk's nest forms the support. Shredded bark underlies the central soft portion of the nest. Surrounding it may be a soft mass of gray lichen and moss. Green live oak twigs, with leaves attached, are often used for the nest's exterior. Tracy I. Storer states that in the Sierra Nevada Western Gray Squirrel nests often are constructed in conifers, while in the coast redwood forests live oaks, California laurel and madrone are used for nest trees. Frequently brood dens are woodpecker or flicker cavities, enlarged by the squirrel occupant.

In late winter or early spring, after a forty-four-day gestation period, single litters of three to five young are produced. If the mother squirrel finds it necessary to retrieve a baby that has tumbled from the nest, she first nudges the baby over, then seizes it by the skin on its chest. The young squirrel then curls its body about its mother's head for the return trip to the nest. When the litter is old enough to be out of the nest and frolic in the den tree, any sign of danger causes the young squirrels instinctively to press against the tree's trunk or a limb.

Late summer and autumn finds Western Gray Squirrels industriously gleaning acorns among the oak tree branches, or gathering them on the ground. Each acorn is carefully buried in a three- to four-inch-deep hole in the forest floor. This habit of burying food in separate holes is a disadvantage to Western Gray Squirrels living in parts of the Sierra Nevada. With excavation of one cache the better-adapted Douglas Squirrel, in the same habitat, uncovers enough food to last for days. Ingles has this to say about the Western Gray Squirrel living at higher elevations:

> The deep snow is probably its worst enemy because of its practice of burying a single fungus or pine nut in one small hole.

In spring these squirrels sometimes strip bark on the upper trunks of young redwood trees, then gnaw away at the underlying succulent layer, a feeding habit that displeases foresters.

Sierra Nevada-dwelling Western Gray Squirrels often attempt to raid caches of the California woodpecker. The squirrels are attracted to their acorn-studded firs or pines—each small shallow cavity in the bark contains an acorn driven in by the woodpecker. With repeated flying attacks the woodpecker threatens to jab the squirrel-thief with its sharply pointed beak and usually succeeds in driving off the trespasser. Seton reports seeing in the Sierra Nevada:

> . . . many dead trees . . . whose holes were plugged with acorns, one in each. These were the caches of the woodpecker; and in defence of them, they waged continuous and noisy warfare on the Gray-squirrels.

Western Gray Squirrels have been accused of raiding orchards, but like other squirrels the good that they do, in assisting in the replanting of trees, outweighs such misdemeanors.

Gray squirrels are not as noisy and excitable as their smaller Douglas Squirrel cousins. Cahalane gives this description of a gray squirrel when annoyed:

> First it barks, a quacking "Yak, yak, yak," repeated many times. Then, as its anger increases, the voice becomes more querulous and higher pitched. The complaint changes to a prolonged harsh "Kuaa-a-a!"

And he adds:

> The bark is not always an impatient one. I have heard it many times when it was soft and had an inquiring or even contented sound. The animal also has quite a repertoire of little chuckles when it is conversing with other squirrels or even itself.

WESTERN GRAY SQUIRREL
(*Sciurus griseus*)

TASSEL-EARED SQUIRRELS

Thousands of years ago tassel-eared squirrels were isolated by formation of the Grand Canyon, and by warmer climates following glacial periods that caused the ponderosa pine forests to flourish at higher elevations. Writing of the Colorado River as a barrier in the distribution of mammals, E. A. Goldman states:

> Perhaps the most strikingly different but obviously allied animals with ranges separated by the Grand Canyon are the Kaibab squirrel (*Sciurus kaibabensis*) and the Abert squirrel (*Sciurus aberti aberti*). The Kaibab squirrel is restricted in range to the yellow pine [ponderosa pine] forest of the Kaibab Plateau, which rises, island-like to 9,000 feet altitude on the northern side. South of the Kaibab Plateau the Granite Gorge of the Colorado is about a mile in depth, and it was in this vicinity that the greatest uplift occurred. Extensive erosion in other directions indicates that the Kaibab squirrel has been isolated since Pleistocene time. The Abert squirrel has an extensive range in yellow pine forest extending to the brink of the Grand Canyon on the southern side.

Presumably both Abert's and the Kaibab squirrels evolved from a common ancestor. But Cahalane writes:

> Their evolution has not gone on for enough thousands of years to standardize racial differences. For instance, an occasional Abert squirrel will turn up wearing a Kaibab black vest, or all-white tail. Sometimes a Kaibab squirrel will appear in a white front and a black tail, apparently borrowed from the Abert race which it has never seen and never will see.

Tassel-eared squirrels are the only North American squirrels that possess conspicuous ear tufts of long hairs, much like those of the tree squirrels of northern Europe and Asia. In summer pelage the ear tufts are missing. By early fall short black hairs project from the ear tips and grow to winter lengths of nearly two inches. During the spring months the tufts fade to brownish and become thin.

Striking color patterns characterize these squirrels. Abert's Squirrel usually sports a grayish tail, white on the underside, and white underparts, while the Kaibab Squirrel's tail is all-white with blackish underparts. The life habits of the two species are very similar.

ABERT'S SQUIRREL (*Sciurus aberti*)

Description. Long ears with tufts of blackish hairs that grow straight up from their tips distinguish Abert's Squirrel. This heavy-bodied, large (eighteen to twenty-one inches long) tree squirrel is gray dorsally, has a reddish brown band running down its back, a blackish lateral band, white underparts and a plume-tail that is grayish above, white beneath and broadly bordered with white hairs. Abert's Squirrels living in central Colorado lack the conspicuous rusty band down the back. Melanistic Abert's Squirrels are common, especially in Colorado.

ABERT'S SQUIRREL
(*Sciurus aberti*)

Distribution. Abert's Squirrels make their homes in coniferous forests (mostly ponderosa pine areas) along the Rocky Mountains in southwestern United States and in Mexico. From the southern rim of the Grand Canyon, Abert's Squirrels range eastward through central Arizona, New Mexico and up into northern Colorado. In Mexico Abert's Squirrels are found in the Sierra Madre Occidental, from northern Sonora and Chihuahua to southern Durango.

Distribution of Abert's Squirrel (top) and the Kaibab Squirrel (bottom)

KAIBAB SQUIRREL (*Sciurus kaibabensis*)

Description. Dark grizzled gray upperparts, usually marked with a rusty dorsal band, dark gray or black underparts and a flashy white tail mark the Kaibab Squirrel. E. A. Goldman writes of the Kaibab or white-tailed squirrel:

> When on the ground, the large waving white tail may, at first glance, give the impression that the animal is a skunk, but its identity is quickly revealed in part by its agility and grace of movement.

Because snowfall is greater and snow cover persists longer into spring at the 9,000-foot elevation of the Kaibab Plateau, it has been suggested that squirrels with whiter tails (held over their backs when feeding) were less conspicuous against a snowy background, and survived to perpetuate their kind, while darker-tailed squirrels were easily detected by predators and removed from the genetic process. But it is more likely that the differences between the Kaibab and Abert's squirrels result from random mutations that became established in the squirrel populations.

Distribution. This squirrel is confined to the Kaibab Plateau on the north rim of the Grand Canyon in northern Arizona, where it has evolved as a product of isolation. The Kaibab Plateau is about 40 miles long and 20 miles wide, bounded on the south by the Canyon; treeless semi-deserts prevent the Kaibab Squirrels from extending their range to the east, north or west. Cahalane writes:

KAIBAB SQUIRREL
(*Sciurus kaibabensis*)

Ruthless hunting or the destruction of the yellow pine [ponderosa] forest could wipe out the entire race of Kaibab squirrels. The southern portion of the plateau is within the Grand Canyon National Park, where no hunting is ever allowed; and the state of Arizona and the United States Forest Service protect the Kaibab squirrels elsewhere. Even museum collectors must wait for accidents, such as highway fatalities, to supply them with a few skins each year.

Considered one of the endangered North American animals, the Kaibab Squirrel is included in *Vanishing Wild Animals of the World.* Richard Fitter states: "The population, like that of other squirrels, fluctuates considerably, and at present, estimated at about 1,000, is thought to be at its lowest ebb for the past fifty years."

According to Natt N. Dodge:

> Over the years, bodies of a number of whitetails have been found without any marks indicating the cause of death. The assumption has been that some disease was, and perhaps still is, afflicting the species. If and what it is and how serious a threat to the population of *kaibabensis* is a matter of conjecture. Some people feel that a continuation of the gradual decline in numbers may eventually result in extermination, while others are confident that the bottom of the population cycle may already be reached and that a gradual buildup of the population will soon begin.

Logging operations and resultant habitat destruction affect the Kaibab Squirrels. Even the automobile, a peril to the squirrels, attests their population decrease. The number of road kills has diminished during the past 30 years, although the number of cars using north rim roads has greatly increased.

From investigations of the Kaibab Squirrel undertaken for the National Park Service, Joseph G. Hall estimates a population of 550 to 1,650 squirrels. About 100 squirrels (plus or minus fifty) live within the Grand Canyon National Park boundaries, while 1,000 (plus or minus 500) are to be found in the Kaibab National Forest, north of the park. The Arizona Game and Fish Commission contests that the present population is the optimum number for the Kaibab Squirrel's limited range and that the species is not endangered.

Habits of the Tassel-eared Squirrels. The tall ponderosa or western yellow pines, between 6,000 and 9,000 feet elevation, are home to the tassel-eared squirrels and provide them with shelter as well as food. Rarely do these squirrels travel down into the oaks below, or into firs above the ponderosa pine zone that is their habitat. Populations of tassel-eared squirrels are generally small. According to Cahalane:

> Sometimes the beautiful big squirrels are conspicuous by their numbers. As many as five or six or even more may be seen chasing each other over a single tree. A year or two later it may be hard to find one. Very likely there is a connection between the squirrel population and the size of the acorn and pine seed crops.

Tassel-eared squirrel nests are usually bulky, about two feet in diameter and constructed of cut pine twigs. They may be situated high in the branches of a pine tree, in a fork, or built upon an old jay's nest. The twigs are worked together to form a rough-walled tight shelter. There may be one, two or three entrances to the nest. Soft grasses and frayed bark form a warm lining. Seton describes a nest of an Arizona tassel-eared squirrel, located fifty feet up in a ponderosa pine:

> Floors, walls and roof, a solid mass of yellow-pine twigs, most of them a foot long and up to ⅜ of an inch in thickness. Outside, it was 18 inches across, and about 12 inches high. The chamber was 7 inches wide and 4 inches high, softly lined with finely shredded fibers, apparently of yucca stalks. It was approached by a tunnel 4 inches in diameter and on the south side. This began under a porch that was 6 inches across. There was also a north exit that was but 2 inches wide. In spite of recent heavy rains, the nest was dry and warm inside; also it was quite clean and without parasites.

Where Abert's Squirrels venture downhill and find abundant hollows in oak trees they sometimes use the tree-holes for their nests. Rarely a large dead pine tree offers a hollow sufficiently large for a squirrel home.

Baby Abert's and Kaibab squirrels are born anytime from early April on, most commonly during the period from May to August. The date depends to some degree upon altitude. When the three or four baby squirrels in the litter are old enough to leave the nest, they follow their mother's example, getting their food from the tall, straight ponderosa pines. Families of Abert's Squirrels sometimes are seen running together on the ground and frolicking in pine trees.

Most active in early morning and late afternoon, tassel-eared squirrels call to each other with soft "whickers." Now and then they bark and scold. E. A. Goldman writes of their feeding habits:

> The white-tailed squirrel gnaws the cones of the yellow pine, but as this tree produces few seeds except in good seed years that come at irregular intervals it must depend largely upon other food. It seems doubtful whether any stores are gathered, the cambium of the pine providing a never-failing supply.

Ponderosa pine bark is the staple in the diet of tassel-eared squirrels. Crawling out almost to the tip of a swaying pine branch, the squirrel bites off the terminal cluster of pine needles and drops it to the forest floor. Then a four- to six-inch piece of the branch is nipped off and carried to a perch where the thick white cambium layer is eaten. The ground below a feeding tree is littered with needles, stripped twigs, branch tips and pieces of gnawed bark.

Leopold describes an Abert's Squirrel in Mexico, gathering acorns:

> After "whickering" in a pine, the squirrel worked its way from limb to limb down to the ground and then started picking acorns from a clump of low scrub oaks one to two feet high. The animal could reach the acorns by sitting up or standing on its hind legs and pulling down the stems with its front paws. It grasped the acorns in its mouth and pulled them from the cups and then buried them. It would gather from one to three acorns, hop a foot or two away, dig a

hole, drop in the nuts, and then cover them, patting the mound down firmly with its front feet and replacing the pine needles over the area.

Seeds of ponderosa pine and other conifers, mushrooms, berries, roots and tubers, seeds of many kinds and green vegetation also are squirrel fare. Sometimes a bird's nest is robbed. According to Edgar A. Mearns:

> The Mexican woodpecker and Abert squirrel have unending and vociferous quarrels and feuds over the latter's propensity for pilfering from the hoards of nuts stored in the pine bark by the busy woodpecker, and scold, berate, and belabor each other soundly.

In 1884 Mearns found Abert's Squirrel "seldom shy . . . quite gentle, and its habits easily observed." In time the squirrels became elusive. Dodge, for several years park ranger on the north rim of the Grand Canyon, writes of some six Kaibab Squirrels that lived near the lodge on Bright Angel Point:

> One or more of these permanent residents might almost always be found and pointed out to visitors who had heard of the rare animals and desired to see one. A particularly bold squirrel frequented the campground and daily visited some of the camp sites soliciting handouts. Unlike the numerous chipmunks and golden-mantled ground squirrels, which became bold enough to accept food from the fingers, the whitetail would rarely approach a person closer than about ten feet.

Any disturbance sends a tassel-eared squirrel dashing up a tree trunk. Keeping the trunk between itself and the danger, the squirrel climbs up to hide motionless among the branches. Like all tree squirrels the tassel-eared squirrel is adept at hiding behind large limbs or trunks where it sits patiently until danger is past; sometimes a squirrel "freezes" for more than an hour at a time. Often, according to E. W. Nelson, "the wind blowing the feathery tip of its tail into view" betrays the squirrel. Nelson describes the leap of a frightened Abert's Squirrel:

> . . . the Squirrel promptly ran to the end of a large branch about fifty feet from the ground, and although no tree was anywhere near on that side, leaped straight out into the air, with its legs outspread just as in a Flying-squirrel. It came down in a horizontal position and struck the ground flat on its under side, and the rebound raised it several inches. Without an instant's delay, it was running at full speed across a little open park, and disappeared in the forest . . .

Of tassel-eared squirrels on the ground, Nelson writes:

> . . . the tail is usually carried upraised in graceful curves. . . . When they walk, they have an awkward waddling gait; but when they are alarmed or desire to move more rapidly for any cause, they progress in a series of extremely graceful bounds, which show the plumelike tail to good advantage. When the Kaibab

> Squirrel is moving about on the ground, its great white tail is extraordinarily conspicuous in the sunshine.

And Cahalane observes:

> Against the dull yellowish brown needles of the summer woodland the white tail of the Kaibab squirrel is conspicuous. Numerous times I have watched the long, gracefully waving plume wandering erratically along between the golden brown tree-shafts while the gray body was almost lost in the shadows.

Enemies of the tassel-eared squirrels include hawks, especially goshawks and red-tailed hawks, coyotes, foxes, bobcats and possibly cougars.

During the winter tassel-eared squirrels avoid storms or extreme cold by remaining in their nests, sometimes for as long as a week or ten days at a stretch. Hoarding is not a conspicuous habit of tassel-eared squirrels. Mostly they depend on ponderosa pine bark, but a few acorns, pine, spruce or fir cones may be recovered from the forest floor litter. Mearns believes that "stores preserved by the provident *carpentero*, or Mexican woodpecker, . . . appropriated . . . whenever found," sustain the squirrels "when the ground is deeply covered with snow and the pine cones mostly fallen or divested of their seeds."

FOX SQUIRREL AND ITS RELATIVES

FOX SQUIRREL (*Sciurus niger*)

Description. Largest of the North American tree squirrels, the heavy-bodied Fox Squirrel has a long, very bushy tail and comes in three main color phases: 1) rusty or reddish gray upperparts, underparts rusty yellow or orange; 2) gray, or gray with some rust coloration on the limbs, blackish on the head, white on nose and ears; 3) black or melanistic. Fox Squirrels in the southeast frequently are melanistic, black or dark brown with white nose and ears. Individuals of the same subspecies exhibit a wide range of variation in color.

Larger size (nineteen to twenty-eight inches total length), a squarer facial profile and four (rather than five) cheek teeth on each side of the upper jaws distinguish the Fox Squirrel and its six close relatives from the gray squirrels.

Distribution. Fox Squirrels once ranged over approximately the eastern half of the United States. Seton notes, in 1928, that Fox Squirrels are "probably exterminated in New England and New York; very rare in New Jersey and Pennsylvania." No longer found in the northeast, Fox Squirrels have extended their range into northern Michigan and westward to include North Dakota, eastern Colorado and Texas. Across the Rio Grande, in northern Mexico, *Sciurus niger* occurs in the states of Coahuila, Nuevo Leon and Tamaulipas.

Introduced into California and other western states Fox Squirrels thrive in many city parks, on large estates and on campuses. Often tame, the squirrels visit bird-feeding stations and learn to accept food from human hands. On the University of

California's Berkeley campus a Fox Squirrel reared its young in a nest of eucalyptus bark in the crotch of a blue gum tree. Fruit of this eucalyptus provides small black seeds that the squirrels relish. Like the squirrels, the eucalyptus trees are introduced species. Ingles reports that around Ventura, California, Fox Squirrels have extended their range into the surrounding countryside. Fox Squirrels also are found in the Seattle area as well as in other parts of Washington.

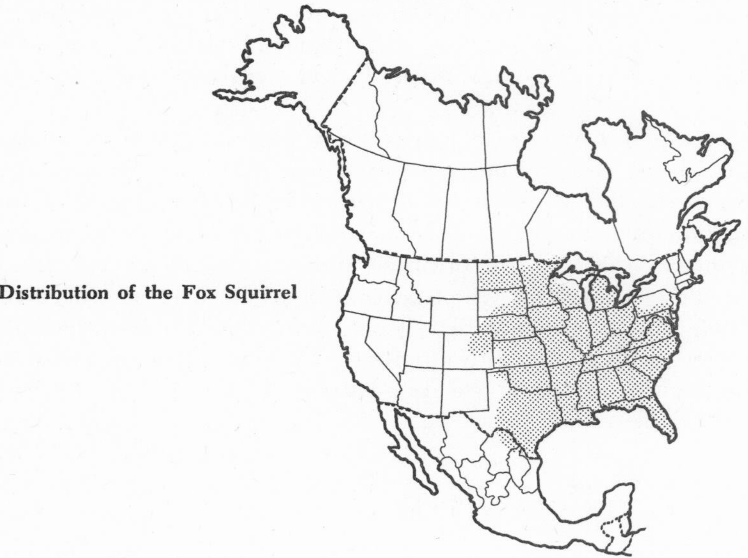

Distribution of the Fox Squirrel

Habits. Open stands or groves of mixed hardwoods are the preferred habitat of most Fox Squirrels. In the northern and western parts of their range Fox Squirrels live on high ridges or well-drained valley bottoms, where oak and hickory trees predominate. Of the Fox Squirrel's choice of habitat Seton writes:

> The Fox-squirrel loves neither the low-timber lands of the Gray, nor the deep pine woods of the Red-squirrel. It is by choice an inhabitant of groves of timber interspersed with open country. The oak "islands" of all the Middlewestern States that lie between the forest and the plains, are ideal for the Fox-squirrel, as also are the belts of timber that skirt the streams of the prairies.

Schwartz reports that in Missouri Fox Squirrels inhabit Osage-orange hedge fences, farm woodlots, timbered fence rows and timbered draws.

In the southeast Fox Squirrels make their homes in open stands of long-leaf pines, or live oaks, sunlit borders of cypress swamps and mangrove swamps. Joseph C. Moore notes that in Florida tall, open forests of long-leaf pines, once the habitat of Fox Squirrels, have been extensively logged. The turkey-oaks that have replaced the pines are low and close-growing, a very different kind of habitat. Moore discovered a number of behavioral adaptations in the Florida Fox Squirrels. Lacking tree cavities or other suitable shelters, the squirrels when frightened sometimes took refuge in underground burrows of gopher tortoises. Some squirrels actually nested underground, in chambers reached by hollows that extended down through old stumps.

Home for most Fox Squirrels is a leafy nest in a tree cavity or fork. A large old hollow tree, often a white oak, elm, sycamore or maple, that "stands out alone at a little distance from the surrounding timber" and affords "a clear view of all going on around him" is preferred for a tree-cavity home, according to Kennicott. Fox Squirrels have been observed to be permanent occupants of such tree homes for as long as five or six successive years. Den homes are usually about six inches wide and fourteen to sixteen inches deep. Abandoned ivory-billed and pileated woodpecker holes were once used, but now Fox Squirrels depend on flicker and red-headed woodpecker cavities, or knot holes and rotted tree crotches, working at the wood with their teeth to enlarge the openings.

Fox Squirrel den openings generally are circular, measure about 2.9 by 3.7 inches and occur on tree trunks where a branch has died and broken off. Scar tissue forms at the base of such limbs, tending to decrease the size of larger openings, and the squirrels themselves continually enlarge the smaller openings. This process of den formation in a woodlot makes Fox Squirrel openings easily recognizable. Sometimes oblong den openings are formed at the base of branches that have been broken off or where the tree has been injured by lightning or wind. Proper Fox Squirrel management, according to Luther L. Baumgartner, must include the saving of potential as well as known den trees:

> Potential den trees are frequently . . . "wolf" trees or worthless snags . . . old white oak, elm, maple, sour gum, beech or chestnut trees, which, if cut, are of little value even for firewood.

Sometimes there is competition for a tree cavity, as this account from Audubon and Bachman attests:

> The Wood Duck is frequently a competitor for the same residence, and contests take place. We have generally observed that the tenant that has already deposited its eggs or young is seldom ejected. The male and female duck unite in chasing and beating with their wings any Squirrel that may approach their nests, nor are they idle with their bills and tongues. On the other hand, when the Squirrel has its young in the hole of a tree and is intruded on, it immediately rushes to the hole, enters it, protrudes its head occasionally, and with a low angry bark keeps possession until the intruder wearies of the contest.

A Fox Squirrel may make its own den in a hollow tree by cutting through to the punky interior and carrying leaves into the cavity for a warm nest lining. Crows' nests have been used by some Fox Squirrels; others have been attracted to use wooden flicker boxes for their nests.

Fox Squirrel homes built of leaves may be hastily constructed summer nests of loosely woven green leaves and twigs, located in a tree fork some twenty feet above ground. Most nest-building activity occurs during summer. Nests built in late winter and spring show a criss-crossed framework of bare twigs with dead leaves packed inside to form the nest. Leaf nests that are used year-round are more carefully constructed, each with a central, roofed-over nest cavity and a small, leaf-concealed side entrance. The exterior is formed of interlaced twigs with leaves attached. Layers of damp leaves are pressed against the inside of the spherical nest to make a tight wall. Shredded bark and bits of leaves are used to line the six- to eight-inch-in-diameter nest cavity. A tree that has a Fox Squirrel nest located anywhere from ten to thirty feet up in its branches usually is the source of the nest material and twigs used in construction, but occasionally leaves and twigs of other trees are brought in. A Fox Squirrel may have two tree cavity homes, or a tree cavity and a leaf nest home. Tree cavity homes are preferred for winter use and for raising families, but when such shelters are scarce leaf nests are used year-round. Durward L. Allen writes:

> From the ground, most leaf nests look small and flimsy, although a closer examination shows that they are by no means so frail as they appear. On several occasions after a rain I evicted a squirrel and found its nest to be dry and warm. Nests that are not regularly used soon cave in and in wet weather become thoroughly soaked.

Often Fox Squirrels occupy a succession of leaf nests, as noted by Cahalane:

> I once watched a fox squirrel that, in twenty-five minutes, made a nest that appeared to be about half a bushel in size. A squirrel may build a new nest every couple of weeks during hot weather, until a dozen in various stages of disrepair are conspicuous in the treetops.

Perhaps the squirrels find a solution to poor housekeeping and parasites when they move to a new nest.

A Fox Squirrel's home range centers about its nest tree. Schwartz states:

> They may use about 10 acres during any one season, but over a year sometimes cover approximately 40 acres by building nests near a new source of food as it becomes seasonally available.

Bad weather may keep a Fox Squirrel in its nest, or restrict its daily activities to feeding within ten to fifteen yards of home. Although they dislike heavy rains, Fox Squirrels have been observed to sit at the base of a high horizontal limb during a drizzling rainfall. With back to tree trunk, the squirrel uses its upward-curved tail for an umbrella. In good weather one or two acres are ranged over in search of food. Fox Squirrel populations become high in the fall, frequently causing the squirrels to move about. Tagged Fox Squirrels have moved from one to fourteen miles from their original home sites; one squirrel reportedly moved forty miles away. Dry years or failure of nut crops also cause squirrels to move about in search of food.

Fox Squirrels stay on the ground much of the time. Hopping along, with fox-like tail waving, the squirrel forages among the leaf litter, searching, smelling, scratching, turning over leaves and pieces of bark for nuts or seeds. If alarmed by a raccoon, bobcat, fox, dog or hawk, the squirrel bounds off over the ground instead of climbing a tree. Usually its dash follows a circular course so that the squirrel seldom travels more than 1,000 feet from its den. Twelve to fifteen miles an hour is maximum speed over short distances for a Fox Squirrel. Outram Bangs writes of the Fox Squirrel's reluctance to take to the trees:

> Its usual way of escape when chased, is to run along the ground to some stump or log, upon which it climbs, and waits until its pursuer comes too near, when it runs to another place of vantage, and so on. It takes to a tree only as a last resort, and then keeps to the trunk and large branches trying to avoid detection by hiding.

Hiding is an "escape" used by the Fox Squirrel, especially when the intruder is a human being. Moore found that Fox Squirrels in Florida living in areas of small, second-growth long-leaf pines climb the plume-like apical twigs of the trees and hide from view by lying on top of the horizontally bent twigs. Once safety of the home tree is reached the Fox Squirrel tends to disappear into its nest, rather than scold and chatter in the manner of the Red, Douglas and gray squirrels.

In the trees Fox Squirrels run along the branches and leap from limb to limb or sometimes from tree to tree. Their awkward aerial leaps often result in falls. Cahalane suggests that perhaps Fox Squirrels have become "too hefty for the most efficient arboreal life." And he relates a Fox Squirrel's fall he witnessed:

> One morning on the campus of the University of Michigan I saw a fox squirrel lose its hold of the elm twigs about forty feet above a concrete sidewalk. Twisting in the air, the animal righted itself and came down on all four feet. It made only a brief pause to shake its head in my direction, as if to say, "Well, what are *you* gawking at?" Then it turned, galloped back to the tree, and ran up as if nothing had happened.

Compared to gray squirrels, Fox Squirrels rise later, go to bed earlier and are more active during the middle of the day. After a few hours of morning foraging the Fox Squirrel returns to its nest or sprawls on a limb to sun itself. During mid-day there may be a second period of activity, or in hot weather the squirrel may stretch out on the base of a high horizontal limb with its tail spread as a shade above its back. In late afternoon the Fox Squirrel comes out to feed again before retiring for the night. When they are busily occupied with autumn nest building or nut caching, the squirrels sometimes stay out at dusk.

During the mid-winter breeding season males often wander outside their home ranges in search of mates, and pairs of Fox Squirrels are seen racing through the trees. Bachman describes such activity:

> In December and January, the season of sexual intercourse, the male chases the female for hours together on the same tree, running up one side and descending on the other, making at the same time a low guttural noise that scarcely bears any resemblance to the barking which they utter on other occasions.

Usually Fox Squirrels have little to do with each other but pairs may stay together during the mating season.

About forty-five days after mating litters of two to five, commonly three, tiny

FOX SQUIRREL
(*Sciurus niger*)

(four-inch-long), naked, pink Fox Squirrels are born. By the second week the fur is apparent on their bodies; their eyes open between the fortieth and forty-fourth days. If disturbed, the baby Fox Squirrels are moved by their mother to another nest. A week or so after their eyes open the young make their first ventures out of the nest, run about the nest tree and sample the foods their mother eats. Weaning occurs at about ten weeks, and when the squirrels are three to four months old they leave the home nest. Like other squirrels, Fox Squirrels show a "compensatory gain" following severe losses. Small breeding populations produce more young, proportionally, than do large breeding stocks. In the southern part of their range Fox Squirrels often produce two litters in the same year. Fox Squirrel breeding seasons are correlated with abundant food supply. In Florida young born in January and February are conceived when the turkey-oak acorn crop is at its peak. The summer breeding season coincides with the ripening of the long-leaf pine seeds.

Fox Squirrel fare consists of nuts and seeds of trees—hickory nuts, walnuts, butternuts, chestnuts, acorns, beech mast, basswood nuts, seeds of maple, elm and ash, conifer seeds, stones of wild cherries and plums, thornapple pits and berries of various kinds. Although Fox Squirrels forage on the ground they invariably carry food to a nearby log, stump, rock or tree perch for eating. In spring the Fox Squirrel's taste for sap causes it to girdle some maples and oaks, stripping the bark with its chisel-like teeth, licking the sap and feeding on the soft cambium. Mushrooms, insects, beetles and grubs, bulbs and roots, berries, corn, buds and twigs are summer foods for Fox Squirrels. Corn, especially in the milk stage, is irresistible to squirrels. Kennicott gives this description:

> Not only does he . . . eat corn on the stalk, but he gnaws off whole ears, which are never eaten on the ground, but always carried off to the fence, or to a neighbouring tree . . . Sitting up on his hind feet, he tears off the husk and pulls out the kernels with his teeth; and if the corn is in the milk, all the soft part is eaten; but when ripened, he often lays down the ear, and, holding the kernels in his paws, only gnaws out the germ. Along fences, or under trees, situated in or near cornfields . . . husks, cobs and partly eaten ears of corn may frequently be found.

Inveterate hoarders, Fox Squirrels have been observed to cache as many as twenty-five hickory nuts in a half-hour period. Whether cut from the trees or found scattered on the ground after the first frost, the nuts usually are stored one to a hole. This feverish fall activity pockmarks the forest floor with hundreds of little cache holes that contain food for the winter.

During very cold weather Fox Squirrels often remain in their nests for several days—a fat squirrel may go for a week or more of bad weather before leaving its nest to feed. Allen writes:

> In the winter their activity appears to be conditioned by temperature and snow depth. Through a blizzard or a period of extreme cold with deep light snow they are likely to remain in the nest. . . . A little snow on the ground or deep snow packed hard does not impede the activity of squirrels. . . . under ideal conditions food supplies will be sufficient to permit fox squirrels to become fat in

the fall. With such a reserve the animals can remain inactive during periods of intense cold and deep snow.

As to recovery of Fox Squirrel caches, Cahalane found

> that a very high percentage . . . is recovered. Periodic examination of two hundred and fifty-one marked caches of acorns and hickory nuts revealed that about ten per cent were emptied before January 1. By spring, all but two of the caches were empty and only one of these nuts was good.

These squirrels rely on memory to find the general vicinity of a food cache and locate nuts by sense of smell. For this reason nuts buried in moist soil are more readily recovered. Every year numbers of cached nuts sprout and grow to perpetuate the forest.

A mange-like condition, probably associated with infestation of scab mites, sometimes causes loss of hair and emaciation. Fox Squirrels afflicted may succumb to exposure or fall victims of predation. Six years is the longevity record for Fox Squirrels in the wild, though few squirrels probably attain such age. In captivity Fox Squirrels have lived as long as ten years.

ARIZONA GRAY SQUIRREL (*Sciurus arizonensis*)

Description. Grizzled gray upperparts, whitish underparts, a brown or yellowish dorsal band and a tail that is black washed with white above and orange or rusty brown below, bordered with black and fringed with white distinguish the Arizona Gray Squirrel. Although it looks like the Gray Squirrel and its habits are similar, it is grouped with the Fox Squirrel, principally because it lacks the peg premolar on either side of its upper jaws.

Distribution. From central Arizona the species ranges into western New Mexico; it is found in the Santa Catalina, Santa Rita and Huachuca mountains of southern Arizona, and in northeastern Sonora and extreme northwestern Chihuahua, Mexico. Ranging as high as the pine forests, this squirrel is also at home in the oak-studded lower mountain elevations.

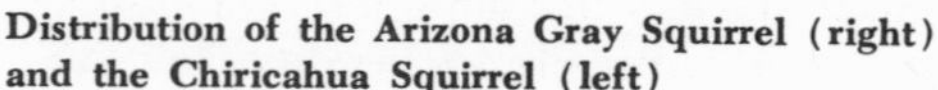

Distribution of the Arizona Gray Squirrel (right) and the Chiricahua Squirrel (left)

CHIRICAHUA SQUIRREL (*Sciurus chiricahuae*)

Description. The Chiricahua Squirrel looks much like a yellowish brown Fox Squirrel. Hall and Kelson state that this squirrel is so closely related to the Apache Squirrel that it may well be only a subspecies of *Sciurus apache.*

Cahalane describes the Chiricahua Squirrel as:

> beautiful but gauche. . . When discovered, instead of dashing off through the trees, this brightly furred animal usually clings to the trunk or branch and tries to "play ostrich." It also likes to spend its leisure hours sunbathing on a branch, dozing and scratching at bothersome mites and lice.

Distribution. As its name suggests this squirrel is found in the Chiricahua Mountains of southeastern Arizona.

Four species of Fox Squirrels are found only in Mexico, in addition to *Sciurus niger,* which ranges into northern Mexico. The Fox Squirrel is known locally as *ardilla* or *ardilla arbórea.* Temperate mountain forests, chiefly pine-oak, are the habitat of almost all Fox Squirrels in Mexico. A. Starker Leopold gives as the range of Mexican Fox Squirrels:

> Bottom lands of the Rio Grande and its principal tributaries in northern Coahuila; northern and central mountains, including the Sierra Madre Occidental, Sierra Madre Oriental, eastern escarpment and southern reaches of the central plateau.

Leopold notes the reduction of their habitat by forest cutting in central Mexico, clearing for agriculture of riverbottom lands in Coahuila, as well as habitat destruction by burning, overgrazing and forest abuse. He concludes:

> In the mountain forests of northern Mexico, however, where the timber has been more recently invaded by man, fox squirrels are still fairly numerous.

Home is a hollow tree den or leaf nest in the treetops. Acorns and other nuts are preferred foods, and these are cached when plentiful, for later excavation. Fruits, seeds, insects and other animal foods, as well as buds and green plant food supplement their diet. Leopold describes Fox Squirrels in Mexico as:

> . . . generally more phlegmatic than their sprightly relatives the gray squirrels . . . They negotiate the treetops adequately but without the spectacular leaps and dashes of their nimble cousins.

In Mexico birth of Fox Squirrel litters occurs during the summer rainy season. According to Leopold:

. . . the timing seems to be keyed to the rainy season rather than to spring. In northwestern Chihuahua we collected two pregnant females of S. *apache* on July 19, one with two embryos of 15 mm. and the other with three of 10 mm. As these young would not have been ready for birth for some weeks, parturition would have occurred in the period of summer rains—a time of plenty in squirreldom. Spring is the period of acute food shortage in the Mexican mountains.

APACHE SQUIRREL (*Sciurus apache*)

Description. Dark grizzled gray upperparts, often lightly washed with yellow or buffy, blackish on crown and mid-back, yellowish or orange underparts, and a tail that is black, washed with yellow above and yellowish rusty to rufous below, bordered with black and fringed with yellow or rufous, are distinguishing characteristics of the Apache Squirrel.

Distribution. Pine-oak forests of the Sierra Madre Occidental in Sonora, Chihuahua, Sinaloa and Durango.

ALLEN'S SQUIRREL (*Sciurus alleni*)

Description. Grizzled yellowish brown above and white below, Allen's Squirrel has a tail that is black tinged with white.

Distribution. Temperate mountain forests of the Sierra Madre Oriental of southeastern Coahuila, Nuevo Leon and Tamaulipas, where it ranges out onto the edges of the Gulf Coastal Plain.

NAYARIT SQUIRREL (*Sciurus nayaritensis*)

Description. Nayarit Squirrels have dark grizzled gray fur, with rusty yellow undertone, and their black tails are conspicuously tinged with white above and rusty below. Feet and underparts are white or yellowish white.

Distribution. Pine and oak forests at 6,500 to 9,000 feet of the western Sierra Madre Occidental in southern Durango, Nayarit, Zacatecas and Jalisco.

PETER'S SQUIRREL (*Sciurus oculatus*)

Description. Sometimes this squirrel has a uniform grizzled gray coat color, or it may have a median band or heavy wash of black. White or buff ear patches and eye rings are characteristic of the species. Its tail is blackish above, grayish or yellowish brown below. Its fur below may be whitish, yellowish or a rich yellowish buff color. Although it clearly resembles the gray squirrel, it lacks the peg premolar and is placed in the subgenus *Parasciurus.*

Distribution. Pine and oak forests of central Mexico, including the states of San Luis Potosi, Guanajuato, Queretaro, Michoacan, Hidalgo, Distrito Federal, Puebla and Veracruz.

TWO SQUIRRELS FROM CENTRAL AMERICA

The subgenus *Guerlinguetus* includes the Tropical Red Squirrel and Richmond's Squirrel.

TROPICAL RED SQUIRREL (*Sciurus granatensis*)

Description. Yellowish or rusty brown hairs intermingle with black on the upperparts of this squirrel; the underparts are a dull rusty buff to deep ferruginous, sometimes with white markings. The tail may be deep red above, or black with an overlay of white hairs and sometimes a black tip. Underneath the tail is dark yellowish brown or dark reddish brown.

Distribution. Tropical jungles in Costa Rica and Panama are the haunts of this squirrel. Like its northern relative *Tamiasciurus,* it is very shy, fleeing danger with spectacular tree-to-tree leaps, or running over the forest floor.

RICHMOND'S SQUIRREL (*Sciurus richmondi*)

Description. Yellowish or rusty brown fur above, yellow or rust-colored underparts, a tail that is black suffused with tawny or rufous hairs above and yellowish brown below, and dull yellow or buff underparts identify Richmond's Squirrel. Skulls of this squirrel, and of the Tropical Red Squirrel, look much like those of *Tamiasciurus.*

Distribution. Tropical forests in central Nicaragua.

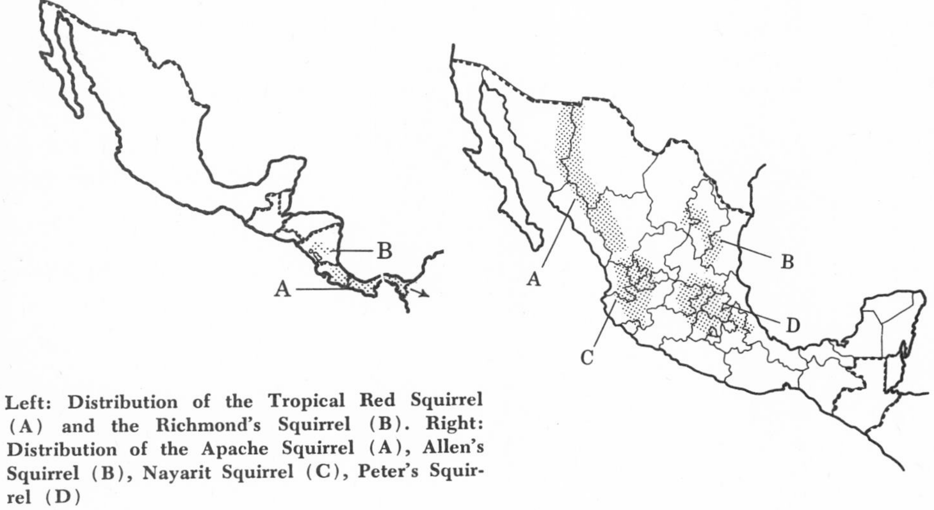

Left: Distribution of the Tropical Red Squirrel (A) and the Richmond's Squirrel (B). Right: Distribution of the Apache Squirrel (A), Allen's Squirrel (B), Nayarit Squirrel (C), Peter's Squirrel (D)

NEOTROPICAL MONTANE GROOVE-TOOTHED SQUIRRELS

Woolly fur, short, broad, densely haired ears, short whiskers and a round, bushy tail characterize the two species of montane squirrels. Closely resembling the dwarf squirrels, montane squirrels are considered by some taxonomists to represent a subgenus rather than a genus. Hall and Kelson state:

> This group, regardless of its nomenclatural status, seems to occupy a position intermediate between the genus *Sciurus* and the genus *Microsciurus*.

BOQUETE MOUNTAIN SQUIRREL
(*Syntheosciurus brochus*)

BOQUETE MOUNTAIN SQUIRREL (*Syntheosciurus brochus*)

This thirteen-inch-long squirrel is tawny olive above and has rust-colored underparts. It has been collected in the Cordillera about eight miles north of Boquete, Chiriqui, Panama.

POÁS MOUNTAIN SQUIRREL (*Syntheosciurus poásensis*)

Cinnamon-buff and black hairs intermingle on the head, back and sides of this squirrel. The fur is darkest down the middle of the back. The tail has a reddish buff fringe. Known from a single specimen collected at Volcán Poás, Alajuela, Costa Rica, this squirrel differs from *Syntheosciurus brochus* by lacking grooves on its upper incisors.

NEOTROPICAL DWARF SQUIRRELS

Three species of *Microsciurus* represent the smallest of the North American squirrels; other species belonging to this genus occur in South America. These squirrels are also called Pygmy Squirrels and, in Mexico, *ardillitas.* L. E. Miller, quoted in J. A. Allen (*Review of South American Sciuridae*) remarks that in Colombia squirrels of this genus are:

> . . . much rarer than other squirrels, and usually [occur] in pairs. They seem to prefer the palm forests that are so abundant on the hillsides where they feed on the various kinds of palm fruits and nuts. They . . . can be approached to within a short distance before taking fright and hiding in the palm leaves. They move rapidly and gracefully, making long, daring leaps.

Small, rounded ears and a relatively narrow tail, shorter than head and body length, characterize the dwarf squirrels.

ALFARO'S DWARF SQUIRREL (*Microsciurus alfari*)

This squirrel, nine to ten inches long, may have olive brown, olive black or yellowish tawny mixed with black for its coat color. Below, its fur varies from buff to yellowish gray. When living at higher elevations its fur is long, dense and soft; at lower elevations its coat is short-haired. Alfaro's Dwarf Squirrel is found in Nicaragua, Costa Rica and Panama.

BOQUETE DWARF SQUIRREL (*Microsciurus boquetensis*)

Probably only a subspecies of *Microsciurus alfari,* this squirrel from Boquete Mountain, Chiriqui, Panama is olive brown in color.

ISTHMIAN DWARF SQUIRREL (*Microsciurus isthmius*)

Grizzled black and buff fur above, orange buffy underparts and a black tail tip distinguish this squirrel, which ranges from South America into eastern Panama.

NEOTROPICAL DWARF SQUIRREL
(*Microsciurus sp.*)

RED SQUIRREL AND DOUGLAS SQUIRREL

RED SQUIRREL (*Tamiasciurus hudsonicus*)

Description. This woodland sentinel wears a rusty red coat, trimmed in summer with black, and a white vest. In winter the Red Squirrel's pelage appears more drab, except for a broad rufous dorsal band, and the fur is noticeably thicker, longer and softer; the reddish ear tufts are more apparent. Loud chatters, barks and sputters, tail flicking and stamping of feet are characteristic of the Red Squirrel. The flattened bushy tail is marked by a broad tapering band of black that gives way to a yellowish fringe. The eyes are ringed with white fur. Smallest of the tree squirrels, Red Squirrels measure eleven to fourteen inches in total length and weigh no more than five to eleven ounces. "The Red-squirrel is a veritable Puck-o'-the-Pines—an embodiment of merriment, bird-like activity and saucy roguery . . . as boisterous as he is vigorous in work and play." So writes Ernest Thompson Seton. Other common names include chickadee, boomer, barking squirrel, pine squirrel, rusty squirrel, red robber, egg-eater, chatterbox and *adjidaumo,* an Indian name meaning "tail-in-air."

Distribution. Red Squirrels lend life and color to much of forested North America. From Alaska and Canada their range extends southward in the Rocky Mountains to New Mexico and in the Appalachians to South Carolina.

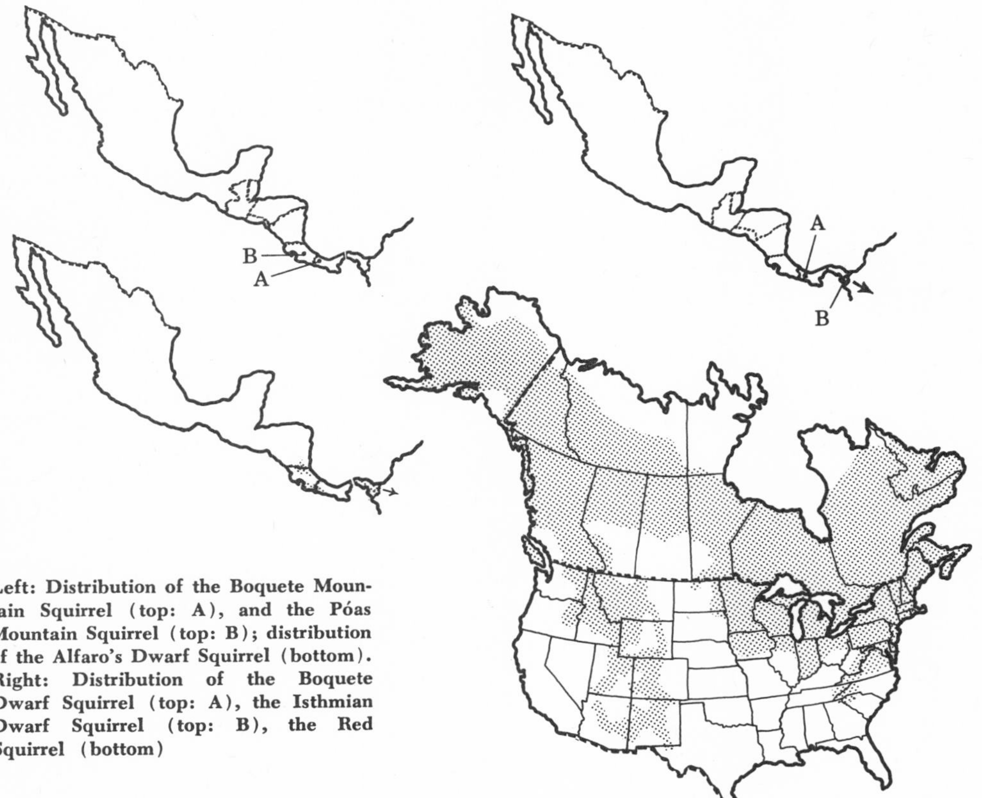

Left: Distribution of the Boquete Mountain Squirrel (top: A), and the Póas Mountain Squirrel (top: B); distribution of the Alfaro's Dwarf Squirrel (bottom). Right: Distribution of the Boquete Dwarf Squirrel (top: A), the Isthmian Dwarf Squirrel (top: B), the Red Squirrel (bottom)

Habits. "Tcher-r-r-r-r" sounds in the late-afternoon silence of the forest. "Tcher-r-r-r-r" comes a reply from a distant tree. Such chatter marks the home territories of the noisy forest sentinels, the Red Squirrels, whose long vibrant calls are carried for various lengths of time and often repeated.

Like other tree squirrels, the Red Squirrel is well equipped for its way of life. Relatively short and narrow as tree-squirrel tails go, the Red Squirrel's tail is nevertheless used for balance in climbing and leaping and lessens the impacts of landings and occasional falls. The tail has other uses, too. With angry flicks it emphasizes a Red Squirrel's fury when an intruder enters its domain. And in cold weather the tail serves as a warm wrapper. Well-developed forelimbs bear slender fingers and vestigial thumbs. Toe pads aid in grasping; thin, curved, sharply pointed claws are useful for climbing, digging in loose soil, rotted wood or snow, as well as for holding food and fur combing. Strong hind limbs and long slender hind feet propel the Red Squirrel's lithe but muscular body up-tree and down with amazing agility and speed. To go up-tree, fore- and hind feet are used alternately in pairs. Coming down, the feet are used individually, unless the squirrel is in a hurry when it descends in a series of jerks, accompanied by falling pieces of bark.

On the ground, where it spends much time foraging, the Red Squirrel walks, makes springing leaps or gallops. Its longest recorded jump on the forest floor is about five feet. When alarmed on the ground a Red Squirrel will flee up a tree, out along a branch to the next tree, down it to the ground, along a runway and up the next tree. By making such use of its three-dimensional habitat the Red Squirrel usually manages to elude even its dread enemy, the marten. So specialized for pursuit of squirrels in their own environment is the marten that Red Squirrels make long forced leaps from tree to tree—their legs extended, bodies flattened, tails held stiff and straight behind—to escape. Being smaller and lighter, the squirrels can project themselves from the ends of branches while the heavier marten must take off from a more secure part of the branch and make a longer leap. Familiarity with all the pathways of its three-dimensional home territory is a life-or-death matter to a Red Squirrel. Sometimes a swaying branch provides an insecure take-off and the Red Squirrel desperately tries to maintain its direction by mid-air contortions of tail and legs. In leaping from tree to tree Red Squirrels may jump as far as eight feet with a drop of two feet, aided on the downglide by kiting action of their small spread bodies and outstretched limbs and tail. When there is no other way to escape, the Red Squirrel makes a forced leap from a high branch to the ground. In this lowest level of its home territory the Red Squirrel maintains a series of runways worn into the forest floor. If the squirrel is a coniferous forest dweller, it also has a labyrinth under and through its midden. Other enemies of the Red Squirrel include fisher, mink, weasels, lynx and bobcat, coyote, fox, domestic cat and various hawks and owls.

Red Squirrel habitats are varied. White spruce provides food and shelter in Alaska's subarctic taiga. In the Rocky Mountain region spruce, fir and lodgepole pine cones are consumed by Red Squirrels. Hemlock groves, larch swamps, bur oak groves and mixed woods of pine, cedar and birch form Midwestern habitats. New England Red Squirrels depend mainly on the white pine, which furnishes cone supply, home site and building materials. White pines shelter Red Squirrels from weather and serve as drying racks for their mushrooms; the tree branches form aerial highways.

Wherever it is found, the Red Squirrel demands an abundant food supply within

RED SQUIRREL
(*Tamiasciurus hudsonicus*)

its home area. The food supply may consist of a great abundance of one kind of food, as in the vast stretches of coniferous forest where Red Squirrels flourish without competition from their cousins, the gray and Fox squirrels. In such coniferous forest habitats Red Squirrels carry on their important work as reforesters—spruce, fir and pine seeds are safely and deeply planted, numbers of them never to be recovered by the squirrels. Or the Red Squirrel may thrive where there is a large variety of foods in different seasons. Mixed stands of conifers and hardwoods offer a varied diet—in spring buds and bark; in summer fruits, such as hazelnuts and blueberries, and mushrooms; and in autumn other fruits and a new crop of cones and nuts. Probably every tree species within a Red Squirrel's home territory is used to some extent, either for food or as a source of nesting material. Trees that are food sources include: white pine, pitch pine, jack pine, red pine, Scotch pine, larch, white spruce, red spruce, black spruce, Norway spruce, Engelmann spruce, Douglas fir, balsam, hemlock, arbor vitae, willow, quaking aspen, balsam poplar, butternut, black walnut, mockernut, pignut, hophorn beam, birch, beech, chestnut, oaks, elms, pear, apple, cherry, sugar maple and other maples and ash. Herbs and fungi also provide Red Squirrel food. Fruits of wild gooseberry, black and red elderberry, raspberry, bunchberry, wild-strawberry and wild rose are sometimes eaten. Red Squirrels are either tolerant or immune to mushroom poison. The many mushroom species that are eaten, fresh or after drying and storage, include the poisonous amanitas. Occasionally Red Squirrels raid eggs in nests or prey upon nestling birds. Like many rodents they relish meat. A carcass lying in the woods provides good gnawing. Red Squirrels may drag a bone or antler to their middens. They have been reported to kill and eat the young of cottontail rabbits and gray squirrels. Snails and insects, including many that are tree pests, also form part of the Red Squirrel's diet. Loose bark is pulled from trees to get at moth pupae and larvae of bark beetles and wood borers. In the southern parts of its range the Red Squirrel competes for food with gray squirrel, Fox Squirrel and flying squirrel. W. E. Cram writes:

> He has a standing grudge against the larger gray squirrel, whom he evidently looks upon as a trespasser . . . you may often see the red chasing the gray round and round and up and down the trunk of hickory or butternut in the full of the year. Yet . . . if the gray squirrel cornered at end of a limb . . . the red squirrel contents himself with backing away to a safe distance and taking it out in scolding.

Merriam states that Red Squirrels contest in early fall with flying squirrels in "nocturnal nut-husking exploits." Nelson also reports nocturnal nut-gathering by Red Squirrels. Such favored foods as hazelnuts are shared with chipmunks and white-footed mice feed on many seeds that Red Squirrels also eat. Mice sometimes raid the Red Squirrel's caches.

Food gathering and storing occupies much of the Red Squirrel's time. Fruits such as those of elm and maple are nipped off, with branch tips and leaves, and dropped to the ground. Cones or nuts are harvested many at a time. Vernon Bailey writes of Red Squirrels in the west:

> Before the seeds are fully matured in the cones they begin to serve as food for the Squirrels, and when well ripened the cones are cut from the pine, spruce

> and fir trees in such numbers that the woods often resound with their steady thumping on ground and logs. During autumn great numbers of cones are cut off and stored in little pockets or holes in the ground, under logs, rocks, or brush-heaps, or in the piles of old cone scales at the base of the feeding trees, where they can be readily found under the deep snows of winter.

Of the Red Squirrel custom of cutting cones before they are dry or ripened Thoreau points out ". . . Their design if I may so speak, in cutting them off green, is partly to prevent their opening and losing their seeds. . ." W. E. Cram states:

> It is the red squirrel who has learned that by cutting off the pine cones in late summer, and burying them under the pine needles, they will ripen and retain their seeds to serve him for food in the winter, months after all the other cones have opened wide their scales and scattered their seeds to the winds.

Harvested cones and nuts then receive careful scrutiny before storage. By licking or moving a food item to another spot the Red Squirrel seems to mark it as its own. Twigs and leaves and even beads of pitch are removed. Red Squirrel forepaws often become stuck with pitch, and the squirrel interrupts its work to remove the pitch with incisor teeth and tongue. If the food is a cone and is to be eaten, the Red Squirrel usually carries it, with head held high, to a branch or stump feeding perch. Holding the cone in its forepaws, the squirrel frees a scale with a swift nip, discards the bract with a jerk of its head and a flip of its forepaw and eats the seed. This process almost always proceeds from base to tip of cone. It takes approximately two minutes for a Red Squirrel to strip a cone; as many as nine cones can be stripped of their seeds in a little less than half an hour. The average number of seeds in a cone is about forty-five. Twelve cones form a Red Squirrel meal, a total of some 540 seeds.

Middens, found throughout Red Squirrel coniferous forest haunts, are formed by husks, shells and scales that accumulate at the bases of stumps, logs or below favorite feeding-perch branches. Their size, usually measured in bushels, depends on species of conifer, the proportion of Red Squirrel diet the seeds compose, the age of the forest and its squirrel population and the number of squirrels that have used the feeding place. Some middens measure twenty feet long, twelve feet wide and three feet in depth and often contain complex networks of tunnels and well-insulated nest cavities.

So strong is the Red Squirrel's instinct for food storage that nearly all leftovers are carefully cached away. According to A. B. Klugh:

> The Squirrel practices three methods of storage: carrying things off to one of its main hoards; burying each object separately, and arranging them in a fork or on a limb. It classifies things that it stores into *two* classes—hard objects and soft Hard things, such as nuts and seeds, it either carries to one of its hoards or buries; soft things such as meat, apples, etc., it arranges about the tree.

In Alaska Olaus J. Murie discovered a Red Squirrel that lived in an abandoned cabin. Mushrooms and other squirrel treasures had been placed for safe-keeping along rustic shelves and in chinks of the log cabin. Food caches, several bushels of seeds and fruits, were piled on and under the cabin bunk. The squirrel's own nest was lined with caribou fur.

Nuts that are gathered green are temporarily cached beneath forest floor litter. The Red Squirrel finds it easier to relocate such food stores after the thick outer coverings have opened. Then these foods are stored in large underground vaults or in hollow logs and stumps. To dig cache pits the Red Squirrel works in dog-like fashion with its forefeet while bracing with its hind feet. The cache is then tamped with the front paws. Mushrooms are gathered one at a time, carried up a tree and carefully placed in forked twigs. Sometimes the branches of a tree will hold several dozen drying mushrooms belonging to a Red Squirrel mycophage. Or the mushrooms may be wedged under loose bark of a tree. After several days or a week the dried mushrooms are added to the squirrel's winter hoard, cached away in hollow trees or underground holes. Such caches furnish the Red Squirrel's food supply during winter and early spring. Red Squirrels have keen memories for their cache locations.

After spring thaw there are leaf buds, flower buds, flowing sap and bark. Branches broken by winter storms provide flowing sap. The Red Squirrel favors sugar maple and hangs by its hind feet to lap the sweet liquid. Hal Borland describes walking in a sugar orchard with a Connecticut farmer one mid-February day:

> All the way he kept looking at the trees, watching the birds and the squirrels, the little red squirrels especially, and the chickadees. . . . Red squirrels know when the sap is about to begin to run. They have a sweet tooth, and at a proper time they nip off twigs which will create taps at which they can drink the first flowing sap. The chickadees follow the red squirrels, for they too like maple sap and they use the taps the squirrels create.

Cram writes that Red Squirrels "gnaw saucer-shaped cavities in the upper side of a branch and drink the sap which fills them." After such a sticky meal, Red Squirrels lick their forepaws and rub them over their noses. Soon flowers, leaves, insects, eggs and nestling birds supplement their diet. Summer and fall staples include nuts, fruits, cones and mushrooms. In fall the Red Squirrels increase their storing activities, preparing for another winter.

Robert T. Hatt, in his study of the Red Squirrel, writes: "The necessity for laying up stores for the winter has developed a sense of ownership and independence in the species which makes it a solitary creature throughout most of the year." Brave and self-reliant, the Red Squirrel possesses a strong property instinct. The safety advantage of its small range, 200 to 250 yards in diameter, is obvious—the Red Squirrel knows all the pathways of escape, through the branches as well as on and underneath the forest floor, and the location of every hole and hollow stump.

Aggressive about its food stores and its home range, the Red Squirrel is fearless in the trees. An intruder is loudly scolded with angry chattering and an occasional explosive "chuck." The Red Squirrel often loses his temper to such an extent that he stamps his hind feet and flicks his tail. Territories are jealously guarded. What appears to be a game of tag is more apt to be a territorial Red Squirrel in pursuit of a squirrel caught poaching. Different intruders evoke different reactions. Enos A. Mills observed a Red Squirrel drop the cone it was carrying and scamper silently up a tree when an owl flew into its territory. But at the sight of a coyote, the squirrel ". . . exploded with a spluttering rush of Squirrel words." The coyote continued on his way but the squirrel ". . . hanging to the cone in his right hand, waved it about and cussed the Coyote as far as he could see him." Often the aggressive Red Squirrel rushes head first down the tree threatening to attack an intruder, then scampers back up the tree, chattering still more excitedly.

Curious by nature, Red Squirrels frequently pause to listen, folding one or both forepaws against their chests. John Burroughs attributes a sense of humor to the Red Squirrel:

> The appearance of anything unusual, if after contemplating it a moment, he concludes it not dangerous, excites his unbounded mirth and ridicule, and he snickers and chatters, hardly able to contain himself; now darting up the trunk of a tree and squealing in derision, then hopping into position on a limb and dancing to the music of his own cackle, and all for your special benefit.

The "tcher-r-r-r" of Red Squirrels is a form of communication—a warning of property rights to other squirrels. If there is no intruder to scold, a Red Squirrel sometimes keeps up a cheerful chatter as it works. Seton describes a singing Red Squirrel: "She seems to amuse herself by uttering all the squirrel notes in rapid succession, going over the list a number of times, and in various combinations, until her performance has lasted ten or fifteen minutes." Seton believes the Red Squirrel's call, passed on from Red Squirrel to Red Squirrel in a coniferous forest, is intended only to frighten the intruder and not to spread alarm among the forest inhabitants. But most observers regard Red Squirrels as forest sentinels. In any case, Red Squirrels are acutely aware of all that goes on about them, whether it is the flapping flight of a raven or a lynx bounding over the snow. Many other animals react to Red Squirrel alarm calls—even the moose of the Alaskan taiga.

For their homes Red Squirrels construct three kinds of nests—tree nests, ground nests and outside nests. Tree cavity nests are favored. Red Squirrels prefer to rear their young in abandoned woodpecker or flicker holes in hollow trees and stubs. The nest holes are usually filled with soft dry grass or lichens, suitable squirrel bedding. Seton writes of a Red Squirrel that shared its nest cavity in a hollow oak with a saw-whet owl. In many forest habitats flying squirrels compete with Red Squirrels for the available nest holes. Ground nests in old rotten stumps, log piles or stone walls offer refuge. A startled Red Squirrel may flee up a tree, follow an aerial highway and then, a hundred yards away, scamper down and enter a ground nest. Ground nests, usually dug beneath a stump, connect with one or more tunnels. Hatt describes such a nest in a spruce stand in New York state. The chamber, a foot below the ground surface, measured about nine by four inches and contained a nest of dry maple and birch leaves lined with finely shredded yellow birch bark. Two holes led into the nest chamber from the forest floor. A third hole opened into the squirrel's catacombs. Red Squirrel excavations range from honeycombed hummocks in spruce stands and burrows in middens that also contain nest and store rooms, to small pits used only for food storage. Wherever their range lacks hollow trees Red Squirrels construct outside or leaf nests. Often a Red Squirrel has two or three leaf nests as well as an underground burrow in its home territory. Outside nests usually measure about sixteen inches in diameter and ten inches in depth. Sometimes they are built on a platform of twigs, or the squirrel may remodel an abandoned hawk or crow nest.

More often the nest is supported by a whorl of branches or built against the tree's trunk, usually near the crown. Nests have been built in witches' broom, a fungus-caused growth. Outside nests are two-layered, the thick outer covering of coarse material surrounding an inner lining of fine soft material. The opening, located on any side except that near the tree's trunk, is sometimes filled with fine nest material to shut out the cold. At a point in the nest opposite the entrance the nest wall is very thin. This is the squirrel's emergency exit, for use in case it is chased home by such an arboreal enemy as the marten. Whatever the nest site, parasites usually share the Red Squirrel's house. In spite of its meticulous grooming the Red Squirrel is not a good housekeeper. Fleas and mites are common boarders and sometimes ticks attach themselves. Botfly larvae may hatch, from eggs deposited on the Red Squirrel's fur, and burrow into its body.

Pairs of Red Squirrels have been seen working together on construction of a nest. Seton writes:

> I have no conclusive testimony to show whether the sexes truly pair, or simply consort for the time being. I have, however, seen two adults at work building a nest, and this is strong evidence, since it is the rule for the male among polygamous animals to shirk all family responsibilities.

Felix Salten's charming tale of Perri and her devoted Porro comes to mind. Whatever the truth about Red Squirrel customs, there is much chasing about among the branches in late winter. Male Red Squirrels fight, and sometimes a squirrel is injured. In April, May or early June, depending upon the latitude of their habitat, baby Red Squirrels are born after a gestation of about six weeks. Litter size varies from two to seven, five or six being the usual number. The newborn squirrels are blind, pink-skinned and a little over four inches long. In a few days fur covers the tops of their heads and the small tails are fringed with hairs. By the thirteenth day the fur has a reddish tinge, and by sixteen days a dark lateral line is noticeable. Soon the babies' undersides are covered with white fur. If the nestful of baby squirrels is disturbed they squeak and churr. Around the twenty-seventh day the young squirrels have their eyes open. The mother squirrel nurses her babies throughout much of the summer. If she has to move her brood or return a wayward youngster to the nest, the mother squirrel grasps the skin of the infant's belly. The baby squirrel's legs and tail then curl around the mother's neck, facilitating transport. Baby Red Squirrels first venture out of their nest when they are about one-third grown. Klugh describes a family of young Red Squirrels playing in the branches of a tree ". . . without displaying any of the self-confident recklessness of their elders, quick to take alarm at the slightest hint of danger. . . ." Until the young squirrels have learned the ways of the forest they remain with their mother. By late August Red Squirrels are busily at work storing food supplies for winter. Many Red Squirrel families remain together through their first winter, and share the family food caches until spring, but some of the earlier broods scatter and establish their own territories nearby.

In favorable habitats Red Squirrel populations usually number about one squirrel per two to three acres. But populations sometimes increase, probably resulting from years of abundant conifer seed crops and correspondingly high survival and productivity among the Red Squirrels. In times of high density emigrations of Red

Squirrels occur. Water does not deter Red Squirrel dispersal, and there are numerous accounts of aquatic Red Squirrels. Although Beatrix Potter may have exaggerated Squirrel Nutkin's use of his tail for a sail, a Red Squirrel *does* swim with at least part of its tail out of the water, as well as its head and top of shoulders. Soaked and unhappy Red Squirrels, far from shore, have climbed up canoe paddles, and after a few minutes rest plunged overboard to resume their swims. A swimming Red Squirrel that was taken into a canoe in the middle of a Maine lake by Alton S. Pope ". . . seemed chilled and very tired and after a moment settled down on [the canoeist's] knee and sat there while [he] paddled ashore." Red Squirrels have even been reported vigorously swimming in Lake Champlain, where the lake is some seven miles in width, their heads above water, "tails erect and expanded." James Higby, in June of 1877, reported "as many as 50" Red Squirrels swimming north on Big Moose Lake in the Adirondacks. The artist Major Allan Brooks once observed a Red Squirrel carrying its youngster, two-thirds grown, into a swift-flowing mining ditch:

> . . . it swam across, landing in front of me, and climbed the bank to my feet. Here it first caught sight of me, threw off the young one, jumped into the water, swam over, ran up the bank and into the woods. The young one ran up to me, stopped on my chest just below my chin for a second or two, then ran down into the water, swam across, ran up the bank and off into the woods after its mother.

Swimming Red Squirrels sometimes fall victims of such fish as large pickerel and pike. Gulls and snapping turtles also prey upon Red Squirrels that take to water.

Of Red Squirrels and cold weather E. W. Nelson writes: "The most intense cold of the Northern winter does not keep them in their nests more than a day or two at a time." According to William O. Pruitt, Jr., Red Squirrels in the Far North go below when air temperatures fall to twenty-five below zero. From the numbing cold of the subarctic snow world the Red Squirrel enters the warm, dark, somewhat humid world of its midden catacomb and tunnels far below the snow. Throughout their range Red Squirrels stay in their nests during stormy or very cold spells. Once the cold snap breaks they reappear, boisterously proclaim territorial boundaries and eagerly dig out their food stores.

T. Donald Carter writes admiringly of Red Squirrel traits:

> The red squirrel, or chickaree, as he is often called, is such a versatile little fellow that it is very difficult to judge his true character. By many of the country people who know him by sight but depend upon the bigoted statements of a few, he is believed to be a mischievous, thieving tyrant that should be destroyed at sight. No birds or squirrels, they say, are safe in the woods where he resides. This is unfortunate, for he certainly is not the villain so often pictured. That he is no saint I am willing to concede, but there is another side to him that when you really get to know him you will appreciate.
>
> I have found red squirrels to be most entertaining and interesting little creatures. They have much more character than the gray squirrel and greatly surpass this squirrel in agility. The red squirrel is impulsive, impudent, and full of initiative . . .

DOUGLAS SQUIRREL (*Tamiasciurus douglasii*)

Description. Like the Red Squirrel the Douglas Squirrel is small, about fourteen inches in total length, of which four or five inches is taken up by a flattened, bushy tail. In spite of different coloration the Douglas Squirrel's alert, nervous manner belies its relationship to the Red Squirrel. Dusky olive or dark grayish brown upperparts and underparts that vary from yellowish white to deep orange mark the Douglas Squirrel's attire. Along each side of its body a narrow black stripe separates the dark and light coat colors. Whitish tipped hairs fringe the tail. The ears, slightly tufted, become tipped with long brownish black hairs in winter.

Named in honor of David Douglas, a botanist who observed the species near the mouth of the Columbia River about 1825, the Douglas Squirrel is celebrated by John Muir:

> . . . the Douglas . . . leaps and glides in hidden strength, seemingly as independent of common muscles as a mountain stream. He threads the tasseled branches of the pine, stirring their needles like a rustling breeze; now shooting across openings in arrowy lines; now launching in curves, glinting deftly from side to side in sudden zigzags, and swirling in giddy loops and spirals around the knotty trunks; getting into what seem to be the most impossible situations without sense of danger; now on his haunches, now on his head; yet ever graceful, and punctuating his most irrepressible outbursts of energy with little dots and dashes of perfect repose. He is without exception, the wildest animal I ever saw—a fiery, sputtering little bolt of life . . .

Allan Brooks claims the Douglas Squirrel's vocabulary differs from that of its red cousin by having "a very musical chirp" or "soft chirping whistle." Robert T. Orr describes the Douglas Squirrel's most common call as "a series of rapidly repeated notes having the mechanical quality of a muffled alarm clock ringing." And he states ". . . especially when curious but wary, they emit an explosive bark and, . . . when chasing each other, . . . a number of chuckling notes." To John Muir the Douglas Squirrel is:

> . . . the mockingbird of squirrels, pouring forth mixed chatter and song . . . barking like a dog, screaming like a hawk, chirping like a blackbird or a sparrow; while in bluff, audacious noisiness he is a very jay.

Of the Douglas Squirrel in Mexico Leopold has this to say:

> Of all the squirrels, the chickaree is perhaps the noisiest and most obstreperous. It livens the woods with its constant chatter, directed at every intruder in its home area or, in the absence of intruders, at neighboring chickarees. Dashing about from limb to limb or hanging on a tree trunk, head down and tail jerking, the little noisemaker warns the world of all dangers, real or suppositi-tious.

Other common names of the Douglas Squirrel include Douglas chickaree, yellow-breasted pine squirrel, yellow-belly, piney sprite and yellow piney. Its Chinook Indian name is *Ap-poe-poe.*

Distribution. Coniferous forests from southern British Columbia south through the Cascades of Washington and Oregon and through the Coast Ranges and down into the southern Sierra Nevada in California. Douglas Squirrels sometimes live at or above timberline. Where trees are scarce or absent they live in the rockslides. A subspecies, *Tamiasciurus douglasii mearnsi,* occurs in coniferous forests high in the Sierra Laguna and Sierra San Pedro Mártir in northern Baja California.

Distribution of the Douglas Squirrel

THE WAYS OF A DOUGLAS SQUIRREL

On a September day in the High Sierra a Douglas Squirrel awoke early. As the warm rays of sun touched the upper branches of the tall pines and firs, the squirrel stirred, stretched and yawned. After peering cautiously down the fir trunk the squirrel bounded to the end of a branch to add his shrill notes, feet dancing and tail jerking, to the clamor made by some jays. A black bear, head low and swinging, trotted by on the forest trail. It was some time before quiet settled over that part of the forest. Soon the only sound was a woodpecker, drilling for beetles. The squirrel started down the fir trunk, moving downward one set of steps, pausing to bark, then a series of fast steps accompanied by churring sounds, then another slow step and bark, all the way to the forest floor, where he paused for a moment with his hind legs still on the trunk. If danger was sensed—a marten, hawk or owl nearby—the squirrel could have whirled back up the trunk. But the ground appeared to be safe and he sprang along in bounds to the base of a Jeffrey pine. Here the squirrel climbed nearly 150 feet to harvest some of the huge cones. As they were cut loose by the squirrel's scissor-like teeth, the cones, some weighing nearly five pounds, had thumped to earth. After four cones were dropped the squirrel came down to eat one and store the others. The cone to be eaten was dragged to the base of a nearby tree. With his back to the tree's trunk the squirrel felt some protection, and he began deftly to cut away the scales, exposing the nutritious seeds underneath. When he had eaten his fill the squirrel gathered the other big cones and dragged them with difficulty to a cache

DOUGLAS SQUIRREL
(*Tamiasciurus douglasii*)

between two boulders. Then he draped himself over a fallen log for his mid-day nap.

It was early fall, the time when Douglas Squirrels are busy harvesting green cones of sequoias, pines and firs. The Douglas Squirrel woke from his nap with a sense of urgency. He climbed a large fir and nipped away at the small cones. After each bite he flipped the cone free with a jerk of his head. Soon more than twenty cones littered the forest floor. He scurried down the tree and carried the cones to a storage hole in a moist stream bank. Moisture would preserve the cones, keeping them tight and in good condition for as long as two or three years. This, as well as his own fondness for fungi, could be a factor in Douglas Squirrel survival in years of cone crop failure. The Douglas Squirrel's caches of small cones were large, some containing hundreds of cones. In winter, depending on sense of smell to locate the cache, he would make a single burrow down through the deep snow to uncover a feast good for many days.

Foresters have been known to make use of Douglas Squirrel caches to obtain seeds of certain conifers. But the squirrels themselves are expert foresters, planting many conifers by sometimes failing to retrieve some of their stores. John Muir describes the Douglas Squirrel's role in forest ecology:

> Nature has made him master-forester, and committed most of her coniferous crops to his paws. Probably over fifty per cent. of all the cones ripened on the Sierra are cut off and handled by the Douglas alone, and of those of the Big Trees perhaps ninety per cent. pass through his hands; the greater portion is of course stored away for food to last during the winter and spring, but some of them are tucked separately into loosely covered holes, where some of the seeds germinate and become trees.

The Douglas Squirrel's mate was also scampering about gathering cones. Late in June she had given birth to a litter of five in her nest of shredded bark and twigs in an abandoned woodpecker hole some thirty feet up in a dead tree. The half-grown youngsters were playing about on the den tree—they would stay at home throughout the winter, feeding on cones their provident mother cached.

The Douglas Squirrel interrupted his cone-collecting when he spied a grouse dust-bathing in the crumbling needles of the forest floor. As the grouse wriggled her body to sift the dust through her feathers the Douglas Squirrel kept up his teasing from a low branch. Leaping forward and back, flicking his tail angrily, the squirrel scolded and churred his protests. But the sudden scream of a red-tailed hawk soaring in the sky sent the squirrel streaking back to his home fir tree, where he crouched still among its branches as the hawk's shrill call cut through the air several more times.

Soon light would begin to fade in the forest and the Douglas Squirrel would climb into the snug shelter of his nest in the fir tree hole.

III

THE GLIDERS

SOUTHERN FLYING SQUIRREL IN FLIGHT
(*Glaucomys volans*)

Delightful goblins of the night, flying squirrels cannot really fly. They are gliders, and have light flattened bodies and lateral skinfolds that extend between their wrists and ankles. On each wrist a tiny cartilagenous process projects the gliding membrane, or patagium, beyond the outstretched limb. With front and hind legs spread the flying squirrel can sail through the air for considerable distances. Large eyes, very soft brownish gray fur, long vibrissae, a short, slightly upturned nose and a horizontally flattened tail are also flying squirrel features.

Grouped in the subfamily Petauristinae with the various kinds of Old World flying squirrels is *Glaucomys,* the genus of the American flying squirrels. This genus includes two species: the Southern Flying Squirrel, *Glaucomys volans,* and the Northern Flying Squirrel, *Glaucomys sabrinus.* These squirrels sleep during the day and at night are active in the same trees that their larger tree squirrel relatives occupy by day. Flying squirrels are seldom seen unless frightened from their nests; sometimes pounding on hollow stumps or dead tree trunks ousts one or more of these nocturnal squirrels.

Known to the early settlers of this continent by the Indian name *assapanick,* the flying squirrel also was called fairy diddle or glider squirrel. Captain John Smith writes in *The Generall Historie of Virginia* (1624):

> A Small beaste they have, they call Assapanick, but we call them flying Squirrels, because spreading their legs, and so stretching the largenesse of their skins, that they have beene seene to fly 30 or 40 yards.

Writing in 1743 Mark Catesby describes the habits of flying squirrels:

> These squirrels are gregarious, travelling from one Tree to another in companies of ten, or twelve together. When I first saw them, I took them for dead Leaves, blown one Way by the Wind, but was not long so deceived, when I perceived many of them to follow one another in one Direction. They will fly fourscore Yards from one Tree to another. They can not rise in their Flight, nor keep in a horizontal Line, but descend gradually, so that in Proportion to the Distance the Tree, they design to fly to, is from them, so much the higher they mount on the Tree they fly from . . .

From *The Quadrupeds of North America* comes this account by The Reverend John Bachman of flying squirrels in a Pennsylvania meadow filled with large oaks and beech trees:

> Suddenly . . . [a flying squirrel] emerged from its hole and ran up to the top of a tree; another soon followed, and ere long dozens came forth and commenced their graceful flights from some upper branch to lower bough. At times one would be seen darting from the topmost branches of a tall oak, and with wide-extended membranes and outspread tail gliding diagonally through the air, til

> it reached the foot of a tree about fifty yards off, when at the moment we expected to see it strike the earth, it suddenly turned upwards and alighted on the body of the tree. It would then run to the top and once more precipitate itself from the upper branches, and sail back again to the tree it had just left. Crowds of these little creatures joined in these sportive gambols; there could not have been less than two hundred. Scores of them would leave each tree at the same moment, and cross each other, gliding like spirits through the air, seeming to have no other object in view than to indulge a playful propensity. We watched and mused til the last shadows of day had disappeared . . .

Perhaps some of the estimates of flying squirrel numbers and volplaning distances are excessive. Twenty to thirty feet is a more usual gliding distance. But sometimes at twilight, or by moonlight, a flying squirrel may be glimpsed gliding in a gentle downward curve from one tree to another. As Evermann and Clark state:

> There is something ghostlike in this gliding flight . . . There is not only an entire absence of fluttering wings, but perfect silence.

Flying squirrels make charming pets. Sometimes people succeed in hand-taming them. Eagerly the little squirrels come gliding down to an outstretched hand for a taste of peanut butter or other treat. F. H. King writes of his pet flying squirrels:

> I have never known wild animals that became so perfectly familiar and confiding as these young Squirrels did; and they seemed to get far more enjoyment from playing upon my person than in any other place, running in and out of pockets, and between my coat and vest. After the frolic was over, they always esteemed it a great favour if I would allow them to crawl into my vest in front, and go to sleep there . . .

My first flying squirrel pet was one that moved into my Connecticut childhood home. For several weeks walnuts and pecans from a large bowl on a living room table had been found cached beneath chair cushions and in corners of the large sofa. My father, who likes nuts, had been blamed for these caches. Then one day we saw a curtain ripple and a flying squirrel sailed from the top of the curtain to land on a bookshelf across the room.

By far the most special squirrel of my acquaintance was a Northern Flying Squirrel named Skitter. Skitter was picked up by a Siamese cat in the Giant Sequoia country near Mariposa, California. He was less than a month old, for his eyes were tightly shut. The cat relinquished Skitter to its owner, Miss Louise Dews, who bottle-fed the tiny squirrel. A year later Skitter came to live at the California Academy of Sciences, where he took part in several television programs and captivated everyone who saw him. In 1960 Skitter retired to become part of our family. He travelled across the country with us, especially enjoying the pinon nuts we collected for him during a visit to the Grand Canyon. Inquisitive and confiding, Skitter found people he knew to be excellent climbing objects and he used their shoulders or the tops of their heads as convenient launching pads and landing sites. If he tired he usually could find a pocket in which to curl up for a nap. When Skitter died, at nine years

SOUTHERN FLYING SQUIRREL
(*Glaucomys volans*)

of age, we were grief-stricken. It was two years before we took in another flying squirrel boarder—the Southern Flying Squirrel that currently holes up by day in the woodpecker hole-log in our dining room.

Poor swimmers because of their gliding skinfolds, flying squirrels' fondness for water often causes them to be drowned in wells, water buckets and toilets. Flying squirrels have even been known to drown in sap buckets, lured to their death by the sweet sap that flows in early spring.

SOUTHERN FLYING SQUIRREL (*Glaucomys volans*)

Description. Literally translated, the scientific name of this squirrel is "flying gray mouse" (from the Greek words *glaukos,* "gray" and *mys,* "mouse" and the Latin *volans,* "flying"). Smaller than its northern relative, this squirrel measures eight and a half to ten and a quarter inches in length, three or four inches of which is tail. Hairs of the underside are white to their roots in the Southern Flying Squirrel, while those of the northern species are dark gray basally. Roundish head, pert nose, very large eyes rimmed with black fur and coat of brownish gray bordered with black along the patagium complete the Southern Flying Squirrel's appearance.

Distribution. The Southern Flying Squirrel is found in hardwood forests throughout the eastern United States, as far west as Minnesota and eastern Texas and in isolated areas in Mexico and Guatemala. It prefers heavy deciduous timber near water, and seldom occurs where coniferous trees predominate. Over some of its range—the Great Lakes region, parts of New England and in the Appalachian Mountains—the Northern Flying Squirrel also occurs. Smaller size (total length is less than eleven inches) and hairs of underparts that are whitish basally distinguish the southern species.

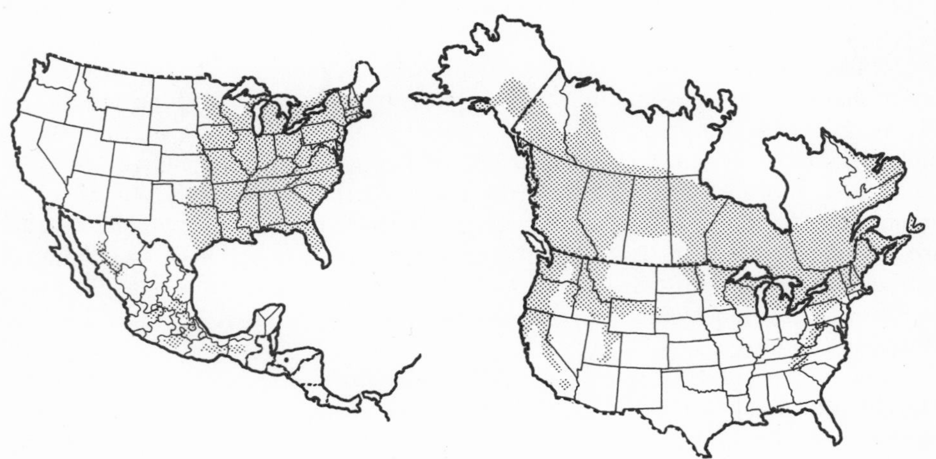

Distribution of the Southern Flying Squirrel (left) and the Northern Flying Squirrel (right)

Habits. Southern Flying Squirrels may inhabit almost any area of scattered hardwoods—beech, maple, oak or hickory—even close to civilization. Old woodlands are preferred, with dead trees and snags showing woodpecker work. As many as fifteen have been found living in the same stump. Bachman quotes Dr. Gideon B. Smith's account of the sociable habits of Southern Flying Squirrels:

> They are gregarious, living together in considerable communities, and do not object to the company of other and even quite different animals. For example, I once assisted in taking down an old martin-box, which had been for a great number of years on the top of a venerable locust tree near my house, and which had some 8 or 10 apartments. As the box fell to the ground, we were surprised to see great numbers of Flying-squirrels, screech-owls, and leather-winged Bats running from it. We caught several of each, and one of the Flying-squirrels was kept as a pet in a cage for six months. The various apartments of the box were stored with hickory nuts, chestnuts, acorns, corn, etc., intended for the winter supply of food. There must have been as many as 20 Flying-squirrels in the box, as many Bats, and we know there were 6 screech-owls.

Flying squirrels also construct their nests of bark shreds, dry leaves, moss, feathers, fur or other soft material in attics and farm buildings, as well as other kinds of bird boxes. Sometimes they build outside or dray nests of twigs and leaves, or take over deserted bird or gray squirrel nests.

Flying squirrel calls include a "chuck, chuck" as well as sharp squeaks and squeals and a bird-like chirping. Merriam has this to say of the flying squirrel at night:

> The Shrew, the Mouse, the Bat, the Chickaree, and the Flying-squirrel are almost sure to be present and the latter is generally responsible for no small share of the perplexing sounds. His activity is intense, his sailing leaps frequent, his gambollings almost ceaseless, his sly chuckle and saucy scold are occasionally heard, and his dropping of beechnut shucks is sometimes well nigh continuous.

Hickory nuts, acorns, sugar maple blossoms, pineberries, various fruits, seeds, fungi, insects, and occasionally birds' eggs or a young bird are food for the Southern Flying Squirrel. Elliptical openings, edged with fine toothmarks, on the sides of hickory nuts or acorns are telltale signs of flying squirrels' presence.

Many accounts of flying squirrel habits state that these squirrels spend little or no time on the ground. In fact, flying squirrels do forage on the forest floor and are readily captured in live traps placed on the ground. At least one flying squirrel, marked by toe-clipping, is known to have traversed a quarter-mile treeless stretch in returning the one and three-quarters miles to her original homesite in a Madison, Wisconsin, house.

The young of this species, usually three or four in a litter, are born during March and April, after a forty-day gestation period. Flying squirrels are devoted mothers, seemingly fearless in the protection of their young. A group of Michigan forestry students once felled a limb containing a flying squirrel nest. The students were curious and gathered around the nest, which one of them picked up and held in his hands. Undaunted, the mother squirrel fled to the top of a nearby tree, volplaned across a river to another tree, climbed up and searched for a suitable hole. Then she glided back across the river, landed near the students and climbed the trouser leg of the fellow holding the nest and her babies. With one baby firmly gripped in her mouth she descended the trouser leg, climbed high up a tree, glided across the river to the new nest-site and poked her baby into the hole. She repeated this feat three times, only once making the mistake of ascending the wrong trouser leg.

Where they are found, flying squirrels are often more numerous than their cousins the Red Squirrels. Favorable habitats may support three to five or more flying squirrels per acre. Their numbers are diminished by reduction of woodland areas and by the practice of removing dead trees with woodpecker nest cavities. As long as their woodland or forest shelter remains, flying squirrel populations hold their own. On predation Paul L. Errington has this to say about the flying squirrel:

> I have wondered if its nocturnal activities and gliding habits might make it vulnerable to the owls out of proportion to its numbers; but it also can be abundant locally and doubtless can have a substantial annual surplus to be frittered away somehow.

Of the flying squirrels' importance to the well-being of a woodland Schwartz writes:

> Their feeding upon the buds of trees probably stimulates better tree growth; their foraging on wood-burrowing insects helps destroy these forms; and their burying of nuts and seeds assists in the continuation of the forest which houses and feeds them. Their body wastes return to the soil as organic matter and add to the sum total contributed by all wildlife. They also serve as a link in the food chain by being preyed upon by other forest dwellers.

NORTHERN FLYING SQUIRREL (*Glaucomys sabrinus*)

Description. In 1801 a zoologist named Shaw first applied the scientific name *Sciurus sabrinus* to a flying squirrel collected near the mouth of the Severn River in Ontario, Canada. The Latin word *sabrina* means a river-nymph, apt reference to the flying squirrel's habit of living near streams and rivers.

Except for its larger size, eleven to twelve inches or more in total length, the Northern Flying Squirrel looks much like the southern species. The sides of the head are gray, the silky pelage of back and sides grayish brown sometimes washed with cinnamon. The tail, broad, flattened and rounded at its tip, is a fuscous gray above and below. Close examination of the whitish undersides reveals that the hairs are slate-colored at their bases.

Distribution. Northern Flying Squirrels are found in old-growth mixed coniferous and deciduous forests from the central part of Alaska across much of Canada to Nova Scotia, south through much of the northeast and in scattered areas in the Appalachian Mountains where the Canadian life zone is bordered by the Transition zone. They are rarely found in hardwood stands, the habitat of the southern species. Their range includes suitable forest areas in the Great Lakes region, Utah, Wyoming, Montana, Idaho, Washington and Oregon. In California Northern Flying Squirrels are found as far south as the San Bernardino and San Jacinto mountains.

THE WAYS OF A NORTHERN FLYING SQUIRREL

A nest of shredded bark and lichens filled the hole abandoned by a hairy woodpecker in the dead snag of a red fir tree. Four tiny flying squirrels, each weighing about a fifth of an ounce and two inches long, with eyes closed, pink, hairless skin and mouse-like tails, were snuggled together in the nest. Their gliding skinfolds were visible, but for the first few days of life the babies were active only to squirm or nurse every two or three hours. Now they were asleep, and their mother had left the nest briefly to feed and scamper about with her mate. The male flying squirrel occupied a nest of his own nearby, constructed of twigs and strips of bark against the trunk of a spruce some twenty feet above the ground. Two holes served as doorways to his nest.

Although it was late May, winter's vestiges remained in patches of snow on the Sierra forest's north-facing slopes. The trees, mostly white pine, lodgepole pine, red fir and mountain hemlock, cast shadows on the forest floor. A creek flowed nearby, its banks lined with mountain alder, willows, aspen, mountain ash, bitter cherry and serviceberry.

During the first weeks of life the baby squirrels remained in the nest. They grew rapidly, and soon their heads were covered with fuzz, their nose whiskers sprouted and their backs turned gray as hair follicles and melanin pigment developed. Soft short fur grew in on their backs. At three weeks their ears opened. At four weeks their eyes opened, but the fully furred young were not ready to leave the nest. As Ernest P. Walker notes:

> Before "flying" squirrels are able to cope with the hazards of great activity in trees and gliding, they must be well developed. "Flying" squirrels are slow in development compared to many terrestrial creatures of similar size.

The baby squirrels climbed about in the nest now. One day two of them ventured out of the nest hole and onto the bark-covered trunk of the fir tree. One youngster tumbled unhurt to the forest floor. The other baby managed to cling to the bark until rescued by its mother. Quick to respond to her babies' squeaks of distress the mother flying squirrel first plucked the baby off the bark, then glided down to the forest floor and with nose and forepaws rolled her baby into a ball. Picking it up by the lateral skinfold of its gliding skinfold, with the baby squirrel's nose and tail curled about her neck, she scurried back up to the nest and poked her offspring through the hole. The mother flying squirrel had two other nests nearby, for use in case she had to move her family to a safer home.

Six weeks after birth the baby flying squirrels began to fend for themselves. Daytime sleeping gave way to nighttime frolics, as their coordination increased. The mother squirrel encouraged them to follow her farther from the nest. The squirrel family's home range, an area of four or five acres, included red firs, lodgepole pines and several large white pines, and numerous stumps and fallen logs of conifers. Because of the flying squirrels' agility aloft it was a three-dimensional range. Where tree branches were continuous the squirrels scampered from tree to tree. More often long leaps or glides were made. Any disturbance sent the youngsters scurrying back to their nest in the fir tree. The little squirrels fed eagerly on the seeds of the coniferous trees and serviceberries. They learned that mushrooms, lichens, beetles and insect larvae make good eating, although they still depended on their mother's milk. One of the young flying squirrels had a narrow escape when his predilection for a carnivorous diet led him to sample the meat bait in a trap set for a pine marten. His mother's alarmed churring warned him just in time.

Emerging from their nest at dusk the four young flying squirrels often could be seen grooming themselves. They used their tongues for washing, their nails for combing and gave special attention to their tails. The fur grew laterally from their tails now, like vanes of a feather.

By their eighth week the squirrels had become proficient gliders and accompanied the mother squirrel on longer trips away from the nest in the fir. Before each glide the youngsters swayed their heads and bodies from side to side, as though gauging their leaps. One by one they sprang and launched downward, to glide like small magic carpets. As each squirrel leaped, with all four legs spread at right angles to its body, the planing surface formed by the patagia was bowed upward by air pressure, much like a miniature parachute. With both skinfolds spread and feather-like tail straight out the flying squirrel glided downward. Glides as long as 160 feet are sometimes made. Over sloping ground a flying squirrel glide of nearly a hundred yards was once recorded.

By means of sharp twists the young flying squirrels learned to avoid branches in their flight paths. Sometimes half jumping, half gliding, they travelled horizontally through the conifers. They were always wary of owls. Large eyes, often characteristic of nocturnal animals, enabled the flying squirrels to gather in enough light for excellent nighttime vision. Sensing danger, the agile gliders could alter flight courses by changing tension in one or both wing skinfolds. For instance, to execute a left turn the left forearm would be dropped, creating aerodynamic drag against the right skinfold and spinning the squirrel into a turn. Sometimes several turns were made in rapid succession, more than once sparing the lives of the young flying squirrels from their nocturnal enemies. In flight the tail, flattened horizontally, functions as rudder and stabilizer.

Downward glides at angles of thirty to fifty degrees, seemed to fill the small clearing. Toward the end of a glide the flying squirrel jerked its stabilizer-tail upward and thrust its feet down so that the camber of its skinfolds formed a tiny parachute, turning its body sharply upward and decreasing speed and jar of landing as hind feet touched down on the tree trunk. Usually they landed with soft thumps low on tree trunks some ten to thirty feet from the take-off sites. Excited high-pitched "chipping" sounds filled the night air and punctuated the noiselessness of their glides. Sometimes emitted at frequencies inaudible to human ears, these shrill noises may have an echo-locating function. Certainly they helped the family of flying squirrels to keep track of each other in their games of aerial tag.

On landing each flying squirrel scrambled to the other side of the tree, its tail appearing to go in the opposite direction. This was an action to foil any pursuing enemy, such as an owl, and was the reason some of the adult flying squirrels had only partial tails. Owls were more to be feared than other flying squirrel enemies—hawks, ravens, lynx and bobcat, raccoon and weasel.

Sometimes the squirrels foraged on the forest floor, but they were less at home on the ground. Hopping in short leaps along fallen logs or over humus-covered soil, they scurried for safety of the nearest tree when alarmed.

NORTHERN FLYING SQUIRRELS
(*Glaucomys sabrinus*)

Once during the summer a family camped for several days near the big red fir. Sitting by their campfire the first evening the people heard soft thuds as the squirrels landed on tree trunks and faint scratchings as they scrambled upward. Delighted that flying squirrels shared their campsite, the people left crumbs of food for them. The squirrels were so confiding that on the second evening the campers could catch glimpses of them gliding from tree to tree in the faint light. After they had gone to bed, the people heard the squirrels skittering about on the sides of their camping tent. Vincent Mowbray gives this account of a Northern Flying Squirrel he caught in a bird trap and then released:

> Several evenings later, at about 8 P.M., a scratching noise on a yellow pine close to the side of the tent attracted my attention to the squirrel which was coming down the tree head first. After watching me for several minutes it came into the tent and ate some stale cornbread muffins. After feeding for about five minutes it took a piece of one of the muffins and went back up the same tree. The next evening . . . presumably the same flying squirrel, repeated the performance . . . This it did every evening for the ensuing two weeks of my stay, always between 8:00 and 8:15.

Autumn was a season of intense activity for the flying squirrel family. The wind blew colder and twilight came earlier. Insects crawled down from the trees to hibernate beneath the litter of the forest floor. The squirrels fed on these as well as on the bounty of tree fruits and seeds, but their favorite foods were fungi and two kinds of lichens: "hair moss" and staghorn lichen. All through the summer they had gathered seeds and nuts, but now their nightly take was greatly increased. Responding to change in their environment as the hours of daylight shortened, the flying squirrels began to eat more each night and to store what they did not eat. Cones were clipped by the squirrels and fell, rattling and bouncing, through the branches to the forest floor. The cones of the red firs, five to eight inches long, turned brownish purple as they ripened, broke open and dropped their seeds. Seeds were collected from the huge white pine cones, each twelve to sixteen inches long. The abundant lodgepole pines' roundish cones bore sharp, recurved spines on their scales, but these did not impede the flying squirrels' harvesting of the seeds. Dismantling a cone, the flying squirrel gnawed on the seed for better hold, then bounded away to hide it. Pulling soil away with its forefeet the squirrel pushed the seed into the ground with its nose, then covered its cache by scattering soil over it with front feet. Large quantities of seeds and nuts were stored away for winter in this manner. Occasionally a seed was missed. It would germinate, become a seedling and perpetuate the coniferous forest. The little squirrels grew noticeably fatter and their dense, soft fur showed traces of the adult pelage. New buffy gray hairs first appeared on the sides, finally spreading over back, head and shoulders. By late November the adult flying squirrels had also completed their yearly change of fur.

Winter brought silence to the Sierra forest. Now and then the strident calls of ravens and jays or the scolding alarm of a Red Squirrel could be heard. Ready now for winter's deepest cold the flying squirrels huddled together in their nest and emerged only for short periods to feed from their stores cached in crevices and crotches of the big fir tree. The five small furry forms, each curled in a ball with its tail over its head, furnished warmth for each other.

That winter the hunting territory of a pine marten included the part of the forest where the flying squirrels nested. Throughout the winter the squirrels eluded the hungry pine marten each time he climbed the big red fir and jumped through its branches. Several flying squirrels in other parts of the forest were less fortunate that winter. The pine marten made the circuit that included the squirrels' fir tree for only a few nights at a time. He travelled the route of another hunting territory before again visiting the big red fir. Come spring he would return to the talus slopes higher up the mountain, where he would hunt pikas, marmots, woodrats and deer mice.

Warm days and cool nights produced buds on some of the forest trees. The flying squirrels found these tiny shoots good to eat, delicately peeling off the outer scale covering of each bud. The conifers put out new cones, bearing their seeds.

The flying squirrels were out of their nest for longer periods at night now. One by one the young left the home in the red fir to search for nest holes of their own. The two male flying squirrels constructed rather flimsy outside nests—twigs and bark strips in branches near nest holes occupied by their new-found mates. One of the nests was actually an old bird's nest which the young male squirrel renovated. The mother squirrel and the two young females prepared their own soft nests for the flying squirrels that would be born later in the spring.

LIST OF REFERENCES

GENERAL

Allen, J. A. "The North American arboreal squirrels," *American Naturalist* 33 (1899), 635–642.

Anthony, H. E. *Field Book of North American Mammals.* New York: G. P. Putnam's Sons, 1928.

Bangs, O. "A review of the squirrels of eastern North America," *Proceedings of the Biological Society of Washington* 10 (1896), 145–167.

Bee, J. W. and E. R. Hall. "Mammals of northern Alaska," *University of Kansas, Museum of Natural History, Miscellaneous Publication* 8 (1956), 35–57.
Observations on the Hoary Marmot and the Arctic Ground Squirrel.

Bourliere, F. *The Natural History of Mammals.* New York: Alfred A. Knopf, 1954.

Brownell, L. W. "How our squirrels pass the winter," *Nature Magazine* 2, No. 5 (1923), 289–291.

Burt, W. H. and R. P. Grossenheider. *A Field Guide to the Mammals.* Boston: Houghton Mifflin Co., 1952.

Cahalane, V. H. *Mammals of North America.* New York: The Macmillan Co., 1947.
Francis Lee Jaques illustrations accompany readable life histories of the 'chucks and ground squirrels, chipmunks and squirrels.

———. *The Imperial Collection of Audubon Animals.* The quadrupeds of North America, original text by J. J. Audubon, F.R.S. and the Rev. John Bachman, D.D. Maplewood, New Jersey: Hammond Inc., 1967.

Crandall, L. S. *The Management of Wild Mammals in Captivity.* Chicago: The University of Chicago Press, 1964.
The care and feeding of many sciurids is detailed.

Errington, P. L. *Of Predation and Life.* Ames: Iowa State University Press, 1967.

Fisher, J., N. Simon and J. Vincent. *Wildlife in Danger.* New York: The Viking Press, 1969.
Kaibab Squirrel, Fox Squirrel (*S. niger cinereus*) and prairie dog (*C. parvidens*) are included.

Goodwin, G. G. *The Mammals of Connecticut.* Hartford: State Geological and Natural History Survey, 1935.

Grinnell, J., J. S. Dixon and J. M. Linsdale. *Fur-bearing Mammals of California.* Berkeley: University of California Press, 1937.

Hall, E. R. *Mammals of Nevada.* Berkeley: University of California Press, 1946.

———. "Handbook of mammals of Kansas," *University of Kansas, Museum of Natural History, Miscellaneous Publication* 7 (1955), 79–102.

——— and K. R. Kelson. *The Mammals of North America.* New York: The Ronald Press Co., 1959.
Descriptions of squirrel species and distribution maps of subspecies are contained in volume one.

———. "Names of species of North American mammals north of Mexico," *University of Kansas, Museum of Natural History,* Miscellaneous Publication 43, 1965, 7–8.

Hamilton, W. J., Jr. *American Mammals, Their Lives, Habits and Economic Relations.* New York: McGraw-Hill Book Co., 1939.

———. *The Mammals of Eastern United States.* Ithaca: Comstock Publishing Co., Inc., 1943.

Ingles, L. G. *Mammals of the Pacific States. California, Oregon and Washington.* Stanford: Stanford University Press, 1965.

Jackson, H. H. T. *Mammals of Wisconsin.* Madison: The University of Wisconsin Press, 1961.

Leopold, A. S. *Wildlife of Mexico.* Berkeley: University of California Press, 1959.
Accounts of gray squirrels, Fox Squirrels, Abert's Squirrel, Deppe's Squirrel and Douglas Squirrel in Mexico.

McClung, R. M. "Squirrels, marmots and prairie dogs," *in Wild Animals of North America.* Washington, D.C.: National Geographic Society, 1960, 230–255.
Illustrations by Louis Agassiz Fuertes and Walter A. Weber, and photographs accompany text.

Mearns, E. A. "Mammals of the Mexican Boundary of the United States," Part 1, *Bulletin of the U.S. National Museum* 56 (1907), 247–349.

Merriam, C. H. *The Mammals of the Adirondack Region.* New York: Henry Holt Co., 1884.

Moore, J. C. "Relationships among the living squirrels of the Sciurinae," *Bulletin of the American Museum of Natural History* 118 (1959), 155–206.

Murie, O. J. *A Field Guide to Animal Tracks.* Boston: Houghton Mifflin Co., 1954.

Orr, R. T. *Mammals of Lake Tahoe.* San Francisco: California Academy of Sciences, 1949.
Excellent accounts of the seven species belonging to the squirrel family that occur in the Lake Tahoe area.

Palmer, R. S. *The Mammal Guide, Mammals of North America North of Mexico.* New York: Doubleday and Co., Inc., 1954.

Rue, L. L., III. *Pictorial Guide to the Mammals of North America.* New York: Thomas Y. Crowell and Co., 1967.

Schwartz, C. W. and E. R. Schwartz. *The Wild Mammals of Missouri.* Kansas City: University of Missouri Press and Missouri Conservation Commission, 1959.
Squirrels found in Missouri are well illustrated and their life histories interestingly detailed.

Seton, E. T. *Lives of Game Animals.* IV, Part 1, Rodents etc. Boston: Charles T. Branford Co., 1953.
Much squirrel lore is contained in this volume first published about 1928.

Shorten, M. *Squirrels.* London: Collins, 1954.

Simpson, G. G. "The principles of classification and a classification of mammals." *Bulletin of the American Museum of Natural History* 85 (1945), 78–80.

Van Gelder, R. G. *Biology of Mammals.* New York: Charles Scribner's Sons, 1969.
Includes discussion of dispersal, home range and territory in squirrels as well as other animals.

Walker, E. P. *et al. Mammals of the World.* Baltimore: The Johns Hopkins Press, 1964.
Volume II contains outline life histories of all the sciurid genera; there is a photograph of each genus.

FOSSIL SCIURIDS AND SCIURID EVOLUTION

Black, C. C. "A review of the North American Tertiary Sciuridae," *Bulletin of the Museum of Comparative Zoology,* Harvard University, 130, No. 3 (1963), 109–248.

Bryant, M. D. "Phylogeny of Nearctic Sciuridae," *American Midland Naturalist* 33, No. 2 (1945), 257–390.

Green, M. "The occurrence of *Citellus richardsonii* (Sabine) in the Pleistocene of Sheridan County, Kansas," *Journal of Mammalogy* 22, No. 4 (1941), 451.

James, G. T. "Paleontology and nonmarine stratigraphy of Cuyama Valley badlands,

California," University of California *Publications in Geological Sciences* 45 (1963), 67–99.

Hibbard, C. W. "*Cynomys ludovicianus ludovicianus* from the Pleistocene of Kansas," *Journal of Mammalogy* 18, No. 4 (1937), 517–518.

Moore, J. C. "The spread of existing diurnal squirrels across the Bering and Panamanian land bridges," *American Museum Novitates* No. 2044 (1961), 1–26.

Stearns, C. E. "A fossil marmot from New Mexico and its climatic significance," *American Journal of Science* 240, No. 12 (1942), 867–878.

Wood, A. E. "Pleistocene prairie-dog from Frederick, Oklahoma," *Journal of Mammalogy* 14, No. 2 (1933), 160.

———. "Eocene radiation and phylogeny of the rodents," *Evolution* 13, No. 3 (1959), 354–361.

WOODCHUCK AND MARMOTS

Abbott, R. L. "Biography of a whistlepig," *Natural History* 43, No. 2 (1939), 112–116.

Armitage, K. B. "Social behavior of a colony of yellow-bellied marmots," *Animal Behavior* 10 (1962), 319–331.

Bailey, V. "The hoary marmot," *National Parks Bulletin* 53 (1927), 9–11.

Beltz, A. and E. S. Booth. "Notes on the burrowing and food habits of the Olympic marmot," *Journal of Mammalogy* 33, No. 4 (1952), 495–496.

Bowdish, B. S. "Tree-climbing woodchucks," *Journal of Mammalogy* 3, No. 4 (1922), 259.

Bronson, F. H. "Daily and seasonal activity patterns in woodchucks," *Journal of Mammalogy* 43, No. 3 (1962), 425–427.

Brown, C. P. "Woodchucks observed while fighting," *Journal of Mammalogy* 29, No. 1 (1948), 70.

Cook, D. B. "Your neighbor the woodchuck," *Audubon* 47, No. 4 (1945), 201–206.

Crisler, L. "The true mountaineer," *Natural History* 59, No. 9 (1950), 422–428.

DeVos, A. and D. Gillespie. "A study of woodchucks on an Ontario farm," *Canadian Field Naturalist* 74, No. 2 (1960), 130–145.

Dowd, C. B. "Rascal, a foundling chuck," *Audubon* 46, No. 4 (1944), 217–220.

Glover, F. A. "A furious woodchuck," *Journal of Mammalogy* 24, No. 3 (1943), 402.

Grizzell, R. A., Jr. "A study of the southern woodchuck," *American Midland Naturalist* 53, No. 2 (1955), 257–293.

Hamilton, W. J., Jr. "The life history of the rufescent woodchuck, *Marmota monax rufescens* Howell," *Annals of the Carnegie Museum* 23 (1934), 85–178.

Hickman, C. P. "Woodchuck climbs trees," *Journal of Mammalogy* 3, No. 4 (1922), 260–261.

Howell, A. H. "Revision of the American marmots," *North American Fauna* 37 (1915), 1–80.

Hoyt, S. F. "Additional notes on the gestation period of the woodchuck," *Journal of Mammalogy* 33, No. 3 (1952), 388–389.

Hoyt, S. Y. "Gestation period of the woodchuck, *Marmota monax*," *Journal of Mammalogy* 31, No. 4 (1950), 454.

Johnson, A. M. "Tree-climbing woodchucks again," *Journal of Mammalogy* 7, No. 2 (1926), 132–133.

Johnson, C. E. "Aquatic habits of the woodchuck," *Journal of Mammalogy* 4, No. 2 (1923), 105–107.

Medsger, O. P. "The tree-climbing habits of woodchucks," *Journal of Mammalogy* 3, No. 4 (1922), 261–262.

Merriam, H. G. "An unusual fox: woodchuck relationship," *Journal of Mammalogy* 44, No. 1 (1963), 115–116.

McHugh, T. "A woodchuck tale," *Explorer,* Cleveland Museum Bulletin 78 (1944), 4–5.

Robb, W. H. "Another tree-climbing woodchuck," *Journal of Mammalogy* 7, No. 2 (1926), 133.

Robinson, W. "Woodchucks and chipmunks," *Journal of Mammalogy* 4, No. 4 (1923), 256–257.

Schoonmaker, W. J. "The value of woodchucks," *Nature Magazine* 27, No. 5 (1936), 302–303.

———. "The woodchuck: lord of the clover field," *New York Zoological Society Bulletin* 41, No. 1 (1938), 3–12.

———. "The woodchuck," *Nature Magazine* 40, No. 1 (1947), 29–32.

———. *The World of the Woodchuck.* Philadelphia: J. B. Lippincott Co., 1966.

Seagears, C. "Wildlife landlord," *New York State Conservationist* 2, No. 5 (1948), 32.

Shilladay, C. L. "Tree-climbing woodchucks," *Journal of Mammalogy* 6, No. 2 (1925), 127–128.

Thorpe, M. R. "A remarkable woodchuck skull," *Journal of Mammalogy* 11, No. 1 (1930), 69–70.

Trump, R. F. "Ways of the woodchuck," *Natural History* 52, No. 5 (1943), 221–225.

Twichell, A. R. "Notes on the southern woodchuck in Missouri," *Journal of Mammalogy* 20, No. 1, (1939), 71–74.

Walker, M. V. "Marmots in the high country," *Yosemite Nature Notes* 25, No. 6 (1946), 88.

Warren, E. R. "Notes on the voice of marmots," *Journal of Mammalogy* 16, No. 2 (1935), 152–153.

PRAIRIE DOGS

Alexander, A. M. "Control not extermination of *Cynomys ludovicianus arizonenesis,*" *Journal of Mammalogy* 13, No. 2 (1932), 302.

Anthony, A. "Behavior patterns in a laboratory colony of prairie dogs, *Cynomys ludovicianus,*" *Journal of Mammalogy* 36, No. 1 (1955), 69–78.

Cates, E. C. "Notes concerning a captive prairie dog," *Journal of Mammalogy* 8, No. 1 (1927), 33–37.

———. "A nature drama," *Journal of Mammalogy* 22, No. 2 (1941), 198–199.

Foster, B. E. "Provision of prairie dog to escape drowning when town is submerged," *Journal of Mammalogy* 5, No. 4 (1924), 266–268.

Garst, W. "Technique for capturing young prairie dogs," *Journal of Wildlife Management* 26, No. 1 (1962), 108.

Hollister, N. "A systematic account of the prairie-dogs," *North American Fauna* 40 (1916), 1–37.

Johnson, G. E. "Observations on young prairie dogs *(Cynomys ludovicianus)* born in the laboratory," *Journal of Mammalogy* 8, No. 2 (1927), 110–115.

Koford, C. B. "The prairie dog of the North American plains and its relations with plants, soil and land use," *Terre et la Vie* 108, Nos. 2–3 (1961), 327–341.

Longhurst, W. "Observations on the ecology of the Gunnison prairie dog in Colorado," *Journal of Mammalogy* 25, No. 1 (1944), 24–36.

Merriam, C. H. "The Prarie-dog of the Great Plains," *Yearbook* U.S. Department of Agriculture (1901), 257–270.

Orcutt, E. "Prairie dogs ways and means," *Zoonooz* 22, No. 10 (1949), 6–8.

Raymond, M. T. and C. O. Mohr. *Prairie Dog Town.* New York: Dodd, Mead and Co. 1942.

Scheffer, T. H. "Study of a small prairie dog town," *Transactions of the Kansas Academy of Science* 40, No. 4 (1938), 391–395.

———. "Ecological comparisons of the plains prairie dog and the Zuni species," *Transactions of the Kansas Academy of Science* 49, No. 4 (1947), 401–406.

Silver, J. "Badger activities in prairie-dog control," *Journal of Mammalogy* 9, No. 1 (1928), 63.

Soper, J. D. "Discovery, habitat and distribution of the black-tailed prairie dog in western Canada," *Journal of Mammalogy* 19, No. 3 (1938), 290–300.

———. "Further data on the black-tailed prairie-dog in western Canada," *Journal of Mammalogy* 25, No. 1 (1944), 47–48.

Stephl, O. E. "Battles between prairie-dog and rattlesnake," *Journal of Mammalogy* 13, No. 1 (1932), 74–75.

Stockard, A. H. "Observations on reproduction in the white-tailed prairie dog," *Journal of Mammalogy* 10, No. 3 (1929), 209–212.

Walker, L. W. "Underground with burrowing owls," *Natural History* 61, No. 2 (1952), 79–95.

Whitehead, L. C. "Notes on prairie dogs," *Journal of Mammalogy* 8, No. 1 (1927), 58.

Young, S. P. "Longevity and other data on a male and female prairie dog kept as pets," *Journal of Mammalogy* 25, No. 3 (1944), 317–319.

GROUND SQUIRRELS

Alcorn, J. R. "Life history notes on the Piute ground squirrel," *Journal of Mammalogy* 21, No. 2 (1940), 160–170.

Baldwin, F. M. and K. L. Johnson. "Effects of hibernation on the rate of oxygen consumption in the thirteen-lined ground squirrel," *Journal of Mammalogy* 22, No. 2 (1941), 180–182.

Balph, D. F. and A. W. Stokes. "On the ethology of a population of Uinta ground squirrels," *American Midland Naturalist* 69, No. 1 (1963), 106–126.

Bartholomew, G. A. and J. W. Hudson. "Desert ground squirrels," *Scientific American* 205 (1961), 107–116.

Blair, W. F. "Rate of development in young spotted ground squirrels," *Journal of Mammalogy* 23, No. 2 (1942), 342–343.

Burt, W. H. "Notes on the habits of the Mohave ground squirrel," *Journal of Mammalogy* 17, No. 3 (1936), 221–224.

Cade, T. "Carnivorous ground squirrels on St. Lawrence Island, Alaska," *Journal of Mammalogy* 32, No. 3 (1951), 358–360.

Cook, A. H. and W. H. Henry. "Texas rock squirrels catch and eat young wild turkeys," *Journal of Mammalogy* 21, No. 1 (1940), 92.

Dubois, A. D. "Hibernating ground squirrel," *Natural History* 39, No. 3 (1937), 214–215.

Edge, E. R. "Seasonal activity and growth in the Douglas ground squirrel," *Journal of Mammalogy* 12, No. 3 (1931), 194–200.

———. "Burrows and burrowing habits of the Douglas ground squirrel," *Journal of Mammalogy* 15, No. 3 (1934), 189–193.

———. "A study of the relation of the Douglas ground squirrel to the vegetation and other ecological factors in western Oregon," *American Midland Naturalist* 16, No. 6 (1935), 949–959.

Emlen, J. T. and B. Glading. "California ground squirrel robs nest of valley quail," *Condor* 15 (1938), 41–42.

Evans, F. C. and R. Holdenreid. "A population study of the Beechey ground squirrel in central California," *Journal of Mammalogy* 24, No. 2 (1943), 231–260.

Evans, F. C. "Notes on a population of the striped ground squirrel in an abandoned field in southeastern Michigan," *Journal of Mammalogy* 32, No. 4 (1951), 437–449.

Gander, F. F. "Development of captive squirrels," *Journal of Mammalogy* 11, No. 3 (1930), 315–317.

Gilmore, R. M. "Notes on an apparent defense attitude in ground squirrels," *Journal of Mammalogy* 15, No. 4 (1934), 322.

Hall, E. R. "A commensal relation of the California quail with the California ground squirrel," *Condor* 29 (1927), 271.

Hansen, R. M. "Microenvironmental influence of periodicity in the Townsend ground squirrel," *Journal of Mammalogy* 37, No. 1 (1956), 124.

Hawbecker, A. C. "Food and moisture requirements of the Nelson antelope ground squirrel," *Journal of Mammalogy* 28, No. 2 (1947), 115–125.

———. "Environmental influence of periodicity in the Townsend ground squirrel," *Journal of Mammalogy* 34, No. 3 (1953) 324–334.

———. "Survival and home range in the Nelson antelope ground squirrel," *Journal of Mammalogy* 39, No. 2 (1958), 207–215.

Hisaw, F. L. and F. E. Emery. "Food selection of ground squirrels, *Citellus tridecemlineatus*," *Journal of Mammalogy* 8, No. 1 (1927), 41–44.

Horn, E. E. "Some relationships of quail and ground squirrels in California," *Journal of Wildlife Management* 2, No. 2 (1938), 58–60.

Howell, A. B. "Food preferences of antelope ground squirrels," *Journal of Mammalogy* 40, No. 3 (1959), 445–446.

Howell, A. B. "Food preferences of antelope ground squirrels," *Journal of Mammalogy* 18, No. 2 (1937), 243–244.

Howell, A. H. "Revision of the North American ground squirrels with a classification of the North American Sciuridae," *North American Fauna* 56 (1938), 1–256.

Ingles, L. G. "Observations of the tree-climbing habits of the California ground squirrel," *Journal of Mammalogy* 26, No. 4 (1946), 438.

Jaeger, E. C. *Desert Wildlife*. Stanford: Stanford University Press, 1961.

Johnson, A. "Notes on the carnivorous propensities of the gray gopher," *Journal of Mammalogy* 3, No. 3 (1922), 187.

Johnson, G. E. "Hibernation of the thirteen-lined ground squirrel, *Citellus tridecemlineatus* (Mitchell); 1. a comparison of the normal and hibernating states," *Journal of Experimental Zoology* 50, No. 1 (1928), 15–30.

———. "Hibernation of the thirteen-lined ground squirrel, *Citellus tridecemlineatus* (Mitchell); II. the general process of waking from hibernation," *American Naturalist* 63, No. 685 (1929), 171–180.

Linsdale, J. M. "Environmental responses of vertebrates in the Great Basin," *American Midland Naturalist* 19, No. 1 (1938), 1–206.

———. *The California Ground Squirrel. A record of observations made on the Hastings Natural History Reservation*. Berkeley: University of California Press, 1946.

Lyon, M. W. "The Franklin ground squirrel and its distribution in Indiana," *American Midland Naturalist* 13, No. 1 (1932), 16–20.

Marsh, R. E. and W. E. Howard. "Breeding ground squirrels, *Spermophilus beecheyi*, in captivity," *Journal of Mammalogy* 49, No. 4, (1968), 781–783.

Martin, R. F. "Occurence of ground squirrels along highways," *Journal of Mammalogy* 16, No. 2 (1935), 154.

Mayer, W. V. "A preliminary study of the Barrow ground squirrel *(Citellus undulatus parryi)*," *Journal of Mammalogy* 34, No. 3 (1953), 334–345.

Miller, L. "A predatory squirrel," *Journal of Mammalogy* 16, No. 4 (1935), 324–325.

Preble, E. A. "A list of the mammals noted on the Seton Expedition of 1907," *in* E. T. Seton's *The Arctic Prairies*. New York: Charles Scribner's Sons, 1911.

Richardson, J. *In* (appendix) *Captain Parry's Journal of a Second Voyage for the Discovery of a Northwest Passage from the Atlantic to the Pacific, Performed in His Majesty's Ships, Fury and Hecla, in the Years 1821–1823 by W. E. Parry*. London: John Murray, 1825.

Rongstad, O. J. "A life history study of thirteen-lined ground squirrels in southern Wisconsin," *Journal of Mammalogy* 46, No. 1 (1965), 76–87.

Shanafelt, M. "*Streak* the story of a ground squirrel and her brood," *Nature Magazine* 16, No. 3 (1930), 155–158.

Shaw, W. T. "Moisture and altitude as factors in determining the seasonal activities of the Townsend ground squirrel in Washington," *Ecology* 2 (1921), 189–192.

———. "The home life of the Columbian ground squirrel," *Canadian Field Naturalist* 38 (1924), 128–130.

———. "The seasonal differences of north and south slopes in controlling activities of Columbian ground squirrels," *Ecology* 6 (1925), 157–162.

———. "Breeding and development of the Columbian ground squirrel," *Journal of Mammalogy* 6, No. 2 (1925), 106–113.

———. "The Columbian ground squirrel as a handler of earth," *Scientific Monthly* 20 (1925), 483–490.

———. "The food of ground squirrels." *American Naturalist* 59, No. 662 (1925), 250–265.

———. "The storing habits of the Columbian ground squirrel," *American Naturalist* 60, No. 669 (1926), 367–373.

———. "Age of the animal and slope of ground surface, factors modifying structure of hibernation dens of ground squirrels," *Journal of Mammalogy* 7, No. 2 (1926), 91–96.

Soper, J. D. "Richardson's ground squirrel," *Nature Magazine* 8, No. 2 (1926), 105–106.

Sowls, L. K. "The Franklin ground squirrel, *Citellus franklinii* (Sabine), and its relationship to nesting ducks," *Journal of Mammalogy* 29, No. 2 (1948), 113–137.

Strumwasser, F. "Some physiological principles governing hibernation in *Citellus beecheyi*," *Bulletin of the Museum of Comparative Zoology*, Harvard University, 124 (1960), 285–320.

Wade, O. "Breeding habits and early life of the 13-striped ground squirrel," *Journal of Mammalogy* 8, No. 4 (1927), 269–276.

———. "The behavior of certain ground squirrels with specific reference to aestivation and hibernation," *Journal of Mammalogy* 11, No. 2 (1930), 160–188.

———. "Soil temperatures, weather conditions, and emergence of ground squirrels from hibernation," *Journal of Mammalogy* 31, No. 2 (1950), 158–161.

GOLDEN-MANTLED GROUND SQUIRREL

Boyer, R. H. "Weasel vs. squirrel in Sequoia National Park," *Journal of Mammalogy* 24, No. 1 (1943), 99–100.

Cameron, D. M., Jr. "Gestation period of the golden-mantled ground squirrel *(Citellus lateralis)*," *Journal of Mammalogy* 48, No. 3 (1967), 492–493.

Follett, W. I. "Prey of weasel and mink," *Journal of Mammalogy* 18, No. 3 (1937), 365.

Hatt, R. T. "Notes on the ground-squirrel *Callospermophilus*," *University of Michigan Occasional Papers*, Museum of Zoology, No. 185, 1927.

———. "The odyssey of a ground squirrel," *Natural History* 29, No. 2 (1929), 181–189.

Killpack, M. L. "A golden-mantled squirrel's race for life," *Journal of Mammalogy* 34, No. 1 (1953), 131.

Selle, R. M. "Golden-mantled ground squirrels raised in captivity," *Journal of Mammalogy* 20, No. 1 (1939), 106–107.

Tevis, L., Jr. "Invasion of a logged area by golden-mantled ground squirrels," *Journal of Mammalogy* 37, No. 2 (1956), 291–292.

CHIPMUNKS

Aldous, S. E. "Food habits of chipmunks," *Journal of Mammalogy* 22, No. 1 (1941), 18–24.

Allen, E. G. "The habits and life history of the eastern chipmunk *(Tamias striatus lysteri),*" *New York State Museum Bulletin* 314, 1938.

———. "Chipmunk secrets," *Animal Kingdom* 46, No. 6 (1943), 133–140.

Anthony, A. W. "Hibernating chipmunks," *Journal of Mammalogy* 5, No. 1 (1924), 76.

Blair, W. F. "Size of home range and notes on life history of the woodland deer mouse and eastern chipmunk in northern Michigan," *Journal of Mammalogy* 23, No. 1 (1942), 27–36.

Broadbrooks, H. E. "Life history and ecology of the chipmunk, *Eutamias amoenus,* in eastern Washington," *Miscellaneous Publications of the Museum of Zoology,* University of Michigan 103 (1958), 5–42.

Brooks, E. A. "A frog-eating chipmunk," *Journal of Mammalogy* 12, No. 3 (1931), 314–315.

Burroughs, J. "Under the apple trees," *Harper's Magazine* 128 (1914), 584–590.

Condrin, J. M. "Observations on the seasonal and reproductive activities of the eastern chipmunk," *Journal of Mammalogy* 17, No. 3 (1936), 231–234.

Crandall, L. S. "A chipmunk kills a sparrow," *Journal of Mammalogy* 17, No. 3 (1936), 287–288.

Damon, D. "Notes on the gray eastern chipmunk," *Journal of Mammalogy* 22, No. 3 (1941), 326–327.

Devoe, A. "The world of the chipmunk," *Audubon* 44, No. 4 (1942), 206–212.

Gordon, K. "The natural history and behavior of the western chipmunk and the mantled ground squirrel," *Oregon State College* (Corvallis) *Studies in Zoology* 5 (1943), 1–104.

Hamilton, W. J., Jr. "Chipmunk," *New York State Conservationist* 4, No. 1 (1949), 14.

Harper, F. "The ways of chipmunks," *Bulletin of the Boston Society of Natural History* 43 (1927), 3–9.

Hostetter, D. R. "Chipmunks and birds," *Journal of Mammalogy* 20, No. 1 (1939), 107–108.

Howell, A. H. "Hibernating habits of chipmunks," *Journal of Mammalogy* 4, No. 2 (1923), 135.

———. "Revision of the American chipmunks," *North American Fauna* 52 (1929). 1–157.

Jameson, E. W., Jr. "Source of food for chipmunks," *Journal of Mammalogy* 24, No. 4 (1943), 500.

Johnson, D. H. "Systematic review of the chipmunks (genus *Eutamias*) of California," *University of California Publications in Zoology* 48 (1943), 63–148.

Klugh, A. B. "Notes on the habits of the chipmunk *Tamias striatus lysteri,*" *Journal of Mammalogy* 4, No. 1 (1923), 29–32.

Miller, A. H. "Specific differences in call notes of chipmunks," *Journal of Mammalogy* 25, No. 1 (1944), 87–89.

Morris, W. A. "The chipmunk as predator of the adult yellow swallowtail butterfly," *Journal of Mammalogy* 34, No. 4 (1953), 510–511.

Nelson, E. W. "Carnivorous habits of the northeastern chipmunk," *Journal of Mammalogy* 16, No. 1 (1935), 66–67.

Panuska, J. A. and N. J. Wade "Captive colonies of *Tamias striatus,*" *Journal of Mammalogy* 41, No. 1 (1960), 122–124.

Robinson, W. "Woodchucks and chipmunks," *Journal of Mammalogy* 4, No. 4 (1923), 256–257.

Schooley, J. P. "A summer breeding season in the eastern chipmunk, *Tamias striatus*," *Journal of Mammalogy* 15, No. 2 (1934), 194–196.

Seidel, D. R. "Homing in the eastern chipmunk," *Journal of Mammalogy* 42, No. 2 (1961), 256–257.

Sherman, A. R. "Periodicity in the calling of a chipmunk," *Journal of Mammalogy* 7, No. 4 (1926), 331–332.

Stevenson, H. M. "Occurrence and habits of the eastern chipmunks in Florida," *Journal of Mammalogy* 43, No. 1 (1962), 110–111.

Test, F. H. "Winter activities of the eastern chipmunk," *Journal of Mammalogy* 13, No. 3 (1932), 278.

Tevis, L., Jr. "Stomach contents of chipmunks and mantled squirrels in northeastern California," *Journal of Mammalogy* 34, No. 3 (1953), 316–324.

Torres, J. K. "A chipmunk captures a mouse," *Journal of Mammalogy* 18, No. 1 (1937), 100.

Walker, A. "A note on the winter habits of *Eutamias townsendi*," *Journal of Mammalogy* 4, No. 4 (1923), 257.

Walker, E. P. "Notes on feeding habits of *Citellus* and *Eutamias*," *Journal of Mammalogy* 26, No. 3 (1945), 308.

White, J. A. "Genera and subgenera of chipmunks," University of Kansas, *Publications of the Museum of Natural History* 5, No. 32 (1953), 543–561.

Yerger, R. W. "Home range, territoriality and populations of the chipmunk in central New York," *Journal of Mammalogy* 34, No. 4 (1953), 448–458.

GRAY SQUIRRELS

Allan, P. F. "Bone cache of a gray squirrel," *Journal of Mammalogy* 16, No. 4 (1935), 326.

Anderson, S. "Tree squirrels (*Sciurus colliaei* group) of western Mexico," *American Museum Novitates* 2093 (1962), 1–13.

Bailey, V. "The Carolina gray squirrel," *Nature Magazine* 5, No. 5 (1925), 303–306.

———. "Gray squirrels," *Frontiers* 5, No. 1 (1940), 13–15.

Baker, R. H. "An ecological study of tree squirrels in eastern Texas," *Journal of Mammalogy* 25, No. 1 (1944), 8–24.

Brooks, F. E. "Note on a feeding habit of the gray squirrel," *Journal of Mammalogy* 4, No. 4 (1923), 257–258.

Clark, A. H. "Speed of gray squirrel," *Journal of Mammalogy* 12, No. 1 (1931), 70.

Dennis, W. "Rejection of wormy nuts by squirrels," *Journal of Mammalogy* 11, No. 2 (1930), 195–201.

Enders, R. K. "The type locality of *Syntheosciurus brochus*," *Journal of Mammalogy* 34, No. 4 (1953), 509.

Fitzwater, W. D., Jr. and W. J. Frank. "Leaf nests of gray squirrels in Connecticut," *Journal of Mammalogy* 25, No. 2 (1944), 160–170.

Flyger, V. F. "Movements and home range of the gray squirrel in two Maryland woodlots," *Ecology* 41, No. 2 (1960), 365–369.

Fussner, B. "Strange behavior of an eastern gray squirrel," *Murrelet* 22, No. 3 (1941), 63.

Habeck, J. R. "Tree-caching behavior in the gray squirrel," *Journal of Mammalogy* 41, No. 1 (1960), 125–126.

Hailman, J. P. "Notes on the following response and other behavior of young gray squirrels," *American Midland Naturalist* 63, No. 2 (1960), 413–417.

Hardy, G. A. "Squirrel cache of fungi," *Canadian Field Naturalist* 63, No. 2 (1949), 86–87.

Harwood, P. D. "Wintering with a gray squirrel," *Audubon* 45, No. 6 (1943), 336–340.

Hibbard, C. W. "Breeding seasons of gray squirrel and flying squirrel," *Journal of Mammalogy* 16, No. 4 (1935), 325–326.

Hungerford, K. E. and N. G. Wilder. "Observations on the homing behavior of the gray squirrel," *Journal of Wildlife Management* 5, No. 4 (1941), 458–460.

Jackson, H. H. T. "A recent migration of the gray squirrel in Wisconsin," *Journal of Mammalogy* 2, No. 2 (1921), 113–114.

Kilham, L. "Gray squirrels born and raised in captivity," *Journal of Mammalogy* 34, No. 4 (1953), 509–510.

Larson, J. S. "Notes on a recent squirrel emigration in New England," *Journal of Mammalogy* 43, No. 2 (1962), 272–273.

Laughlin, F. J. "Depredations of a gray squirrel," *Journal of Mammalogy* 26, No. 4 (1946), 440–441.

Longley, W. H. "Minnesota gray and fox squirrels," *American Midland Naturalist* 69, No. 1 (1963), 82–98.

Nichols, J. T. "Notes on the food habits of the gray squirrel," *Journal of Mammalogy* 8, No. 1 (1927), 55–57.

Petrides, G. A. "A gall insect food of the gray squirrel," *Journal of Mammalogy* 25, No. 4 (1944), 410.

Reilly, E. M., Jr. "Squirrels on the move," *New York State Conservationist* (Dec.–Jan. 1968–69), 6.

Schorger, A. W. "An emigration of squirrels in Wisconsin," *Journal of Mammalogy* 28, No. 4 (1947), 401–403.

Seton, E. T. "Migrations of the gray squirrel," *Journal of Mammalogy* 1, No. 2 (1920), 53–58.

———. "Gray squirrels and nuts," *Journal of Mammalogy* 2, No. 4 (1921), 238–239.

———. *Bannertail, the Story of a Gray Squirrel.* New York: Charles Scribner's Sons, 1922.

Terres, J. K. "Gray squirrel utilization of elm," *Journal of Wildlife Management* 3, No. 4 (1939), 358–359.

Thoma, B. L. and W. H. Marshall. "Squirrel weights and populations in a Minnesota woodlot," *Journal of Mammalogy* 41, No. 2 (1960), 272–273.

Uhlig, H. G. "Gray squirrel populations in extensive forested areas of West Virginia," *Journal of Wildlife Management* 21, No. 3 (1957), 335–341.

Woods, G. T. "Midsummer food of gray squirrels," *Journal of Mammalogy* 22, No. 3 (1941), 321–322.

WESTERN GRAY SQUIRREL

Fritz, E. "Squirrel damage to young redwood trees," *Journal of Mammalogy* 13, No. 1 (1932), 76.

Hatt, R. T. "A gray squirrel carries its young," *Journal of Mammalogy* 8, No. 3 (1927), 244–245.

Ingles, L. G. "Ecology and life history of the California gray squirrel," *California Fish and Game* 33, No. 3 (1947), 139–158.

Merriam, C. H. "A nest of the California gray squirrel," *Journal of Mammalogy* 11, No. 4 (1930), 494.

Ross, R. C. "California Sciuridae in captivity," *Journal of Mammalogy* 11, No. 1 (1930), 76–78.

Storer, T. I. "The young of the California gray squirrel," *Journal of Mammalogy* 3, No. 3 (1922), 188–189.

ABERT'S AND KAIBAB SQUIRRELS

Bailey, F. M. "Abert squirrel burying pine cones," *Journal of Mammalogy* 13, No. 2 (1932), 165–166.

Dodge, N. N. "Whitetail squirrel," *Pacific Discovery* 18, No. 2 (1965), 23–26.

Goldman, E. A. "The Kaibab or white-tailed squirrel," *Journal of Mammalogy* 9, No. 2 (1928), 127–129.

McCartney, E. S. "Calling on the Kaibab squirrel," *Nature Magazine* 29, No. 5 (1937), 271–272.

Thornburg, R. and F. Thornburg. "Tuft-eared squirrels," *Nature Magazine* 39, No. 10 (1946), 523–524.

Wade, O. "Notes on the northern tuft-eared squirrel, *Sciurus aberti* in Colorado," *American Midland Naturalist* 16, No. 2 (1935), 201–202.

FOX SQUIRRELS

Allen, D. L. "Populations and habits of the fox squirrel in Allegan County, Michigan," *American Midland Naturalist* 27, No. 2 (1942), 338–379.

Baumgartner, L. L. "Fox squirrel dens," *Journal of Mammalogy* 20, No. 4 (1939), 456–465.

———. "Fox squirrels in Ohio," *Journal of Wildlife Management* 7, No. 2 (1943), 193–202.

Boulware, J. T. "Eucalyptus tree utilized by fox squirrel in California," *American Midland Naturalist* 26, No. 3 (1941), 696–697.

Brown, L. G. and L. E. Yeager. "Fox squirrels and gray squirrels in Illinois," *Bulletin Illinois Natural History Survey* 23, No. 5 (1945), 449–536.

Canalane, V. H. "Out of season caching by fox squirrel," *Journal of Mammalogy* 11, No. 1 (1930), 78.

———. "Caching and recovery of food by western fox squirrel," *Journal of Wildlife Management* 6, No. 4 (1942), 338–352.

Cottam, C. "How fast can a fox squirrel run?," *Journal of Mammalogy* 22, No. 3 (1941), 323.

Dice, L. R. "How do squirrels find buried nuts?," *Journal of Mammalogy* 8, No. 1 (1927), 55.

Greene, H. C. "A record of fox squirrel longevity," *Journal of Mammalogy* 31, No. 4 (1950), 454–455.

Hicks, E. A. "Ecological factors affecting activity of western fox squirrel," *Ecological Monographs* 19, No. 4 (1949), 287–302.

Moore, J. C. "The natural history of the fox squirrel *(Sciurus niger shermani)*," *Bulletin of the American Museum of Natural History* 113, No. 1 (1957), 1–71.

Sherman, A. R. "Fox squirrels' nests in a barn," *Journal of Mammalogy* 7, No. 4 (1926), 332.

Stoddard, H. L. "Nests of the western fox squirrel," *Journal of Mammalogy* 1, No. 3 (1920), 122–123.

Svihla, R. D. "Captive fox squirrels," *Journal of Mammalogy* 12, No. 2 (1931), 152–156.

RED SQUIRRELS

Blackford, J. L. "Cone year," *Nature Magazine* 39, No. 8 (1946), 409–412.

Borland, H. *Homeland; a Report from the Country*. Philadelphia: J. B. Lippincott Co., 1969, 50–51.

Carter, T. D. "Red squirrel chatter," *Audubon* 61, No. 3 (1959), 110–113.

Caton, J. D. "Mode of drinking of the red squirrel," *American Midland Naturalist* 13, No. 1 (1879), 46.

Clarke, C. H. D. "Some notes on hoarding and territorial behavior of the red squirrel," *Canadian Field Naturalist* 53, No. 3 (1939), 42–43.

Cole, L. J. "Red squirrels swimming a lake," *Journal of Mammalogy* 3, No. 1 (1922), 53-54.

Cram, W. E. "The red squirrel," *Journal of Mammalogy* 5, No. 1 (1924), 37–41.

Finley, W. L. and I. Finley. "The wit of a red squirrel," *Nature Magazine* 5, No. 3 (1925), 141–144.

Gordon, K. "Territorial behavior and social dominance among Sciuridae," *Journal of Mammalogy* 17, No. 2 (1936), 171–172.

Grange, W. B. "Pine squirrel carrying young," *Journal of Mammalogy* 9, No. 2 (1928), 151–152.

Hatt, R. T. "The red squirrel: its life history and habits," *Roosevelt Wildlife Annals* 2, No. 1 (1929), 1–146.

———. "The red squirrel farm," *Natural History* 29, No. 5 (1929), 319–326.

Hayward, C. L. "Feeding habits of the red squirrel," *Journal of Mammalogy* 21, No. 2 (1940), 220.

Ingram, W. M. "Red squirrels chased by robins," *Journal of Mammalogy* 21, No. 2 (1940), 220.

Kilham, L. "Territorial behavior of red squirrels," *Journal of Mammalogy* 35, No. 2 (1954), 252–253.

Klugh, A. B. "Ecology of the red squirrel." *Journal of Mammalogy* 8, No. 1 (1927), 1–32.

Mailliard, J. "Redwood chickaree testing and storing hazel nuts," *Journal of Mammalogy* 12, No. 1 (1931), 68–70.

Murie, O. J. "The Alaska red squirrel providing for winter," *Journal of Mammalogy* 8, No. 1 (1927), 37–40.

Petrides, G. A. "Snow burrows of the red squirrel," *Journal of Mammalogy* 22, No. 4 (1941), 393–394.

Pope, A. S. "Swimming red squirrels," *Journal of Mammalogy* 5, No. 2 (1924), 134.

Preston, F. W. "Red squirrels and gray," *Journal of Mammalogy* 29, No. 3 (1948), 297–298.

Pruitt, W. O., Jr. *Animals of the North.* New York: Harper and Row, Publishers, 1966. The life of a red squirrel, sentinel of the taiga, pages 16-27.

Seagears, C. "The red squirrel," *New York State Conservationist* 4, No. 3 (Dec. 1949–Jan. 1950), 40–44.

Shaw, W. T. "Moisture and its relation to the cone-storing habit of the western pine squirrel," *Journal of Mammalogy* 17, No. 4 (1936), 337–349.

Svihla, R. D. "Development of young red squirrels," *Journal of Mammalogy* 11, No. 1 (1930), 79–80.

Warren, E. R. "Fremont's squirrel kitchen midden," *Journal of Mammalogy* 13, No. 3 (1932), 278.

Yeager, L. E. "Cone-piling by Michigan red squirrels," *Journal of Mammalogy* 18, No. 2 (1937), 191–194.

FLYING SQUIRRELS

Booth, E. S. "Notes on the life history of the flying squirrel," *Journal of Mammalogy* 27, No. 1 (1946), 28–30.

Catesby, M. *The Natural History of Carolina.* London: printed at the expense of the author, 1743, part II, 76–77.

Cowan, I. McT. "Nesting habits of the flying squirrel *Glaucomys sabrinus*," *Journal of Mammalogy* 17, No. 1 (1936), 58–60.

Doutt, J. K. "*Glaucomys sabrinus* in Pennsylvania," *Journal of Mammalogy* 11, No. 2 (1930), 239.

Evermann, B. W. and H. W. Clark. "Notes on the mammals of the Lake Maxinkuckee Region." *Proceedings of the Washington Academy of Sciences* 13, No. 1 (1911), 14–16.

Findley, J. S. "The interesting fate of a flying squirrel," *Journal of Mammalogy* 26, No. 4 (1946), 437.

Fryxell, F. M. "Flying squirrels as city nuisances," *Journal of Mammalogy* 7, No. 2 (1926), 133.

Gordon, D. C. "Adirondack record of flying squirrel above timberline," *Journal of Mammalogy* 43, No. 2 (1962), 262.

Hatt, R. T. "Habits of a young flying squirrel," *Journal of Mammalogy* 12, No. 3 (1931), 233–238.

Howell, A. H. "Revision of the American flying-squirrels," *North American Fauna* 44 (1918), 29–58.

Ingles, E. "Mac—the flying squirrel," *Nature Magazine* 30, No. 4 (1937), 229.

Jordan, J. S. "Notes on a population of eastern flying squirrels," *Journal of Mammalogy* 37, No. 2 (1956), 294–295.

King, F. H. "Instinct and memory exhibited by the flying squirrel in confinement, with a thought on origin of wings in bats," *American Naturalist* 17, No. 1 (1883), 36–42.

Kittredge, J. "Can the flying squirrel count?," *Journal of Mammalogy* 9, No. 3 (1928), 251–252.

Klugh, A. B. "The flying squirrel," *Nature Magazine* 3, No. 4 (1924), 205–207.

MacClintock, D. "Gliders of the night," *Pacific Discovery* 16, No. 1 (1963), 11–15.

Maslowski, K. "The story of Woolly, a flying squirrel," *Nature Magazine* 32, No. 8 (1939), 441–444.

McCabe, R. A. "Homing of flying squirrels," *Journal of Mammalogy* 28, No. 4 (1947), 404.

McKeever, S. "Food of the northern flying squirrel in northeastern California," *Journal of Mammalogy* 41, No. 2 (1960), 270–271.

Morris, P. A. "A flying squirrel mother," *Nature Magazine* 22, No. 4 (1933), 177.

Mowbray, V. "Notes on the Sierra Nevada flying squirrel," *Journal of Mammalogy* 20, No. 3 (1939), 379.

Muul, I. and J. W. Alley. "Night gliders of the woodlands," *Natural History* 72, No. 5 (1963), 18–25.

Muul, I. "Mating behavior, gestation period, and development of *Glaucomys sabrinus*," *Journal of Mammalogy* 50, No. 1 (1969), 121.

Osgood, F. L. "Apparent segregation of sexes in flying squirrels," *Journal of Mammalogy* 16, No. 3 (1935), 231.

Perkins, G. H. "The flying squirrel," *American Naturalist* 7, No. 3 (1873), 132–139.

Robert, E. "Life with a flying squirrel infant," *Nature Magazine* 44, No. 2 (1951), 80.

Smith, J. *The Generall Historie of Virginia*. London: printed by I. D. and I. H. for Michael Sparkes, 1624.

Snyder, L. L. "An outside nest of a flying squirrel." *Journal of Mammalogy* 2, No. 3 (1921), 171.

Sollberger, D. E. "Notes on the life history of the small eastern flying squirrel," *Journal of Mammalogy* 21, No. 3 (1940), 282–293.

———. "Notes on the breeding habits of the eastern flying squirrel (*Glaucomys volans volans*)," *Journal of Mammalogy* 24, No. 2 (1943), 163–173.

Stack, J. W. "Courage shown by a flying squirrel, *Glaucomys volans*," *Journal of Mammalogy* 6, No. 2 (1925), 128.

Stoddard, H. L. "The flying squirrel as a bird killer," *Journal of Mammalogy* 1, No. 2 (1919), 95–96.

Svihla, R. D. "A family of flying squirrels," *Journal of Mammalogy* 11, No. 2 (1930), 211–213.

Walker, E. P. "'Flying' squirrels—nature's gliders," *National Geographic Magazine* 91, No. 5 (1947), 662–674.

———. "They glide through the air with the greatest of ease," *Fauna* 10, No. 3 (1948), 82–84.

———. "Glimpses of flying squirrels," *Nature Magazine* 44, No. 2 (1951), 81–84.

———. "The flying squirrel, nature's glider," In *Wild Animals of North America*. Washington, D.C.: National Geographic Society, 1960, 256–263.

Wood, K. "Sundown squirrel," *Fauna* 3, No. 4 (1941), 107–112.